T0255456

CAMBRIDGE LIBRARY COLLECTION

Books of enduring scholarly value

Botany and Horticulture

Until the nineteenth century, the investigation of natural phenomena, plants and animals was considered either the preserve of elite scholars or a pastime for the leisured upper classes. As increasing academic rigour and systematisation was brought to the study of 'natural history', its subdisciplines were adopted into university curricula, and learned societies (such as the Royal Horticultural Society, founded in 1804) were established to support research in these areas. A related development was strong enthusiasm for exotic garden plants, which resulted in plant collecting expeditions to every corner of the globe, sometimes with tragic consequences. This series includes accounts of some of those expeditions, detailed reference works on the flora of different regions, and practical advice for amateur and professional gardeners.

The Natural History of the Tea-Tree

This treatise on the tea bush and the consumption of tea was published in 1772 by John Coakley Lettsom (1744–1815), a physician and philanthropist, whose first action on inheriting his family plantation in 1767 was to free all its slaves. He practised medicine in London, and wrote on topics which he felt would benefit society. The book begins with a description of the plant, using the Linnaean system, discussing tea cultivation and harvesting in China, and the preparation of the leaves for use locally and abroad. In Part II, Lettsom turns to the medical uses of tea, lamenting that so little scientific evidence exists for either its beneficent or its malign properties. After performing various experiments and considering the physical and social consequences of tea-drinking, he concludes that it should be avoided, because its enervating effects lead to weakness and effeminacy: advice which mostly fell on deaf ears.

The Natural History of the Tea-Tree

With Observations on the Medical Qualities of Tea, and Effects of Tea-Drinking

John Coakley Lettsom

CAMBRIDGE
UNIVERSITY PRESS

University Printing House, Cambridge, CB2 8BS, United Kingdom

Cambridge University Press is part of the University of Cambridge.
It furthers the University's mission by disseminating knowledge in the pursuit of education, learning and research at the highest international levels of excellence.

www.cambridge.org
Information on this title: www.cambridge.org/9781108079815

This edition first published 1772
This digitally printed version 2015

ISBN 978-1-108-07981-5 Paperback

Selected botanical reference works available in the
CAMBRIDGE LIBRARY COLLECTION

al-Shirazi, Noureddeen Mohammed Abdullah (compiler), translated by Francis Gladwin: *Ulfáz Udwiyeh, or the Materia Medica* (1793) [ISBN 9781108056090]

Arber, Agnes: *Herbals: Their Origin and Evolution* (1938) [ISBN 9781108016711]

Arber, Agnes: *Monocotyledons* (1925) [ISBN 9781108013208]

Arber, Agnes: *The Gramineae* (1934) [ISBN 9781108017312]

Arber, Agnes: *Water Plants* (1920) [ISBN 9781108017329]

Bower, F.O.: *The Ferns (Filicales)* (3 vols., 1923–8) [ISBN 9781108013192]

Candolle, Augustin Pyramus de, and Sprengel, Kurt: *Elements of the Philosophy of Plants* (1821) [ISBN 9781108037464]

Cheeseman, Thomas Frederick: *Manual of the New Zealand Flora* (2 vols., 1906) [ISBN 9781108037525]

Cockayne, Leonard: *The Vegetation of New Zealand* (1928) [ISBN 9781108032384]

Cunningham, Robert O.: *Notes on the Natural History of the Strait of Magellan and West Coast of Patagonia* (1871) [ISBN 9781108041850]

Gwynne-Vaughan, Helen: *Fungi* (1922) [ISBN 9781108013215]

Henslow, John Stevens: *A Catalogue of British Plants Arranged According to the Natural System* (1829) [ISBN 9781108061728]

Henslow, John Stevens: *A Dictionary of Botanical Terms* (1856) [ISBN 9781108001311]

Henslow, John Stevens: *Flora of Suffolk* (1860) [ISBN 9781108055673]

Henslow, John Stevens: *The Principles of Descriptive and Physiological Botany* (1835) [ISBN 9781108001861]

Hogg, Robert: *The British Pomology* (1851) [ISBN 9781108039444]

Hooker, Joseph Dalton, and Thomson, Thomas: *Flora Indica* (1855) [ISBN 9781108037495]

Hooker, Joseph Dalton: *Handbook of the New Zealand Flora* (2 vols., 1864–7) [ISBN 9781108030410]

Hooker, William Jackson: *Icones Plantarum* (10 vols., 1837–54) [ISBN 9781108039314]

Hooker, William Jackson: *Kew Gardens* (1858) [ISBN 9781108065450]

Jussieu, Adrien de, edited by J.H. Wilson: *The Elements of Botany* (1849) [ISBN 9781108037310]

Lindley, John: *Flora Medica* (1838) [ISBN 9781108038454]

Müller, Ferdinand von, edited by William Woolls: *Plants of New South Wales* (1885) [ISBN 9781108021050]

Oliver, Daniel: *First Book of Indian Botany* (1869) [ISBN 9781108055628]

Pearson, H.H.W., edited by A.C. Seward: *Gnetales* (1929) [ISBN 9781108013987]

Perring, Franklyn Hugh et al.: *A Flora of Cambridgeshire* (1964) [ISBN 9781108002400]

Sachs, Julius, edited and translated by Alfred Bennett, assisted by W.T. Thiselton Dyer: *A Text-Book of Botany* (1875) [ISBN 9781108038324]

Seward, A.C.: *Fossil Plants* (4 vols., 1898–1919) [ISBN 9781108015998]

Tansley, A.G.: *Types of British Vegetation* (1911) [ISBN 9781108045063]

Traill, Catherine Parr Strickland, illustrated by Agnes FitzGibbon Chamberlin: *Studies of Plant Life in Canada* (1885) [ISBN 9781108033756]

Tristram, Henry Baker: *The Fauna and Flora of Palestine* (1884) [ISBN 9781108042048]

Vogel, Theodore, edited by William Jackson Hooker: *Niger Flora* (1849) [ISBN 9781108030380]

West, G.S.: *Algae* (1916) [ISBN 9781108013222]

Woods, Joseph: *The Tourist's Flora* (1850) [ISBN 9781108062466]

For a complete list of titles in the Cambridge Library Collection please visit:
www.cambridge.org/features/CambridgeLibraryCollection/books.htm

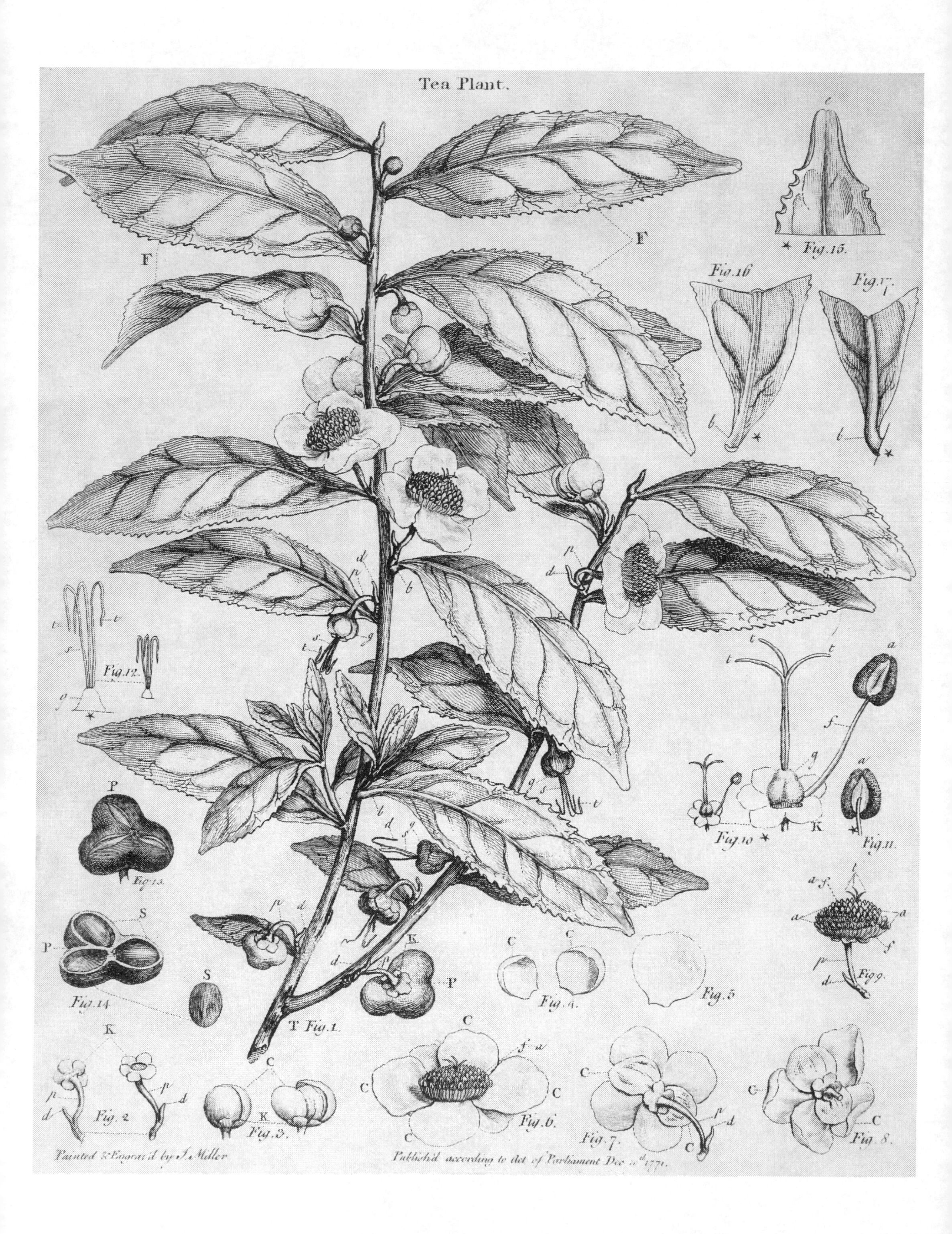
Tea Plant.
Fig. 1.
Fig. 2
Fig. 3
Fig. 4
Fig. 5
Fig. 6.
Fig. 7.
Fig. 8.
Fig. 9.
Fig. 10
Fig. 11.
Fig. 12.
Fig. 13.
Fig. 14
Fig. 15.
Fig. 16
Fig. 17
Painted & Engrav'd by J. Miller
Publish'd according to Act of Parliament Dec 5th 1771.

THE NATURAL HISTORY OF THE TEA-TREE, WITH OBSERVATIONS ON THE MEDICAL QUALITIES of TEA, AND EFFECTS OF TEA-DRINKING.

By JOHN COAKLEY LETTSOM, M. D. F. S. A.

Ad utilitatem vitæ omnia consilia factaque nostra dirigenda sunt. TACIT.

LONDON:
Printed for EDWARD and CHARLES DILLY, in the Poultry.
M DCC LXXII.

TO THE

DUKE

OF

NORTHUMBERLAND,

THE FOLLOWING

TREATISE

IS

WITH PERMISSION,

RESPECTFULLY INSCRIBED

BY

THE AUTHOR.

PREFACE.

THE ſubject of the following eſſay being now in general uſe among the inhabitants of this kingdom, as well as in many other parts of Europe, and conſtituting ſo large a part of commerce, I imagined it would afford no ſmall degree of pleaſure to the curious to have ſome account of the natural hiſtory of a ſhrub, with the leaves of which they are ſo well acquainted.

Many treatiſes have been publiſhed on the uſes and effects of Tea; a few writers have likewiſe given ſome circumſtances relative to its natural hiſtory and preparation, the indefatigable Kæmpfer particularly: but theſe circumſtances lie ſo diſperſed, and the accounts which have been given of the virtues and efficacy of Tea, are in general

ſo

ſo contradictory, and void of true medical obſervation, that it ſeemed no improper ſubject for a candid diſcuſſion. The reader will at leaſt have the ſatisfaction of ſeeing in a narrow compaſs, the principal opinions relative to this ſubject.

Within theſe three or four years we have been ſucceſsful enough to introduce into this kingdom a few genuine Tea plants. There was formerly, I am told, a very large one in England, the property of an Eaſt-India captain, who kept it ſome years, and refuſed to part with either cuttings or layers. This died, and there was not another left in the kingdom. A large plant was not long ſince in the poſſeſſion of the great Linnæus, which, I am informed, is now dead. I know ſeveral gentlemen, who have ſpared neither pains nor expence to procure this evergreen from China, but their beſt endeavours proved unſucceſsful. For though many ſtrong and good plants were ſhipped at Canton, and all poſſible care taken of them during the voyage, yet they ſoon grew ſickly, and but one till of late has ſurvived the paſſage to England.

The largeſt Tea plant in this kingdom, is, I believe, at Kew; it was preſented to that royal ſeminary by J. Ellis, Eſq; who raiſed it from the ſeed.

ſeed. But the plant at Sion-houſe, belonging to the Duke of Northumberland, is the firſt that ever flowered in Europe; and an elegant drawing has been taken from it in that ſtate, with its botanical deſcription. The engraver has done juſtice to his original drawing, which is now in the poſſeſſion of that great promoter of natural hiſtory, Dr. Fothergill, to whom I have been indebted for many dried ſpecimens and flowers of the Tea-tree from China. If the reader compares this plate with the following deſcription, he will have as clear an idea of this exotic ſhrub, as can at preſent be exhibited.

There have been likewiſe a few young Tea plants lately introduced into ſome of the moſt curious botanic gardens about London; ſo that it ſeems probable this very diſtinguiſhed vegetable will become a denizen of England, and ſuch of her colonies as may be deemed moſt favorable to its propagation.

In regard to the effects of Tea on the human conſtitution, one might have imagined that long and general uſe would have furniſhed ſo many indiſputable proofs of its good and bad properties, that nothing could be eaſier than to determine theſe with preciſion: yet ſo difficult a thing it is

to

to eſtabliſh phyſical certainty in regard to the operation of food or medicines on the human body, that our knowledge in general, even with reſpect to this article, is very imperfect. Nevertheleſs, I have endeavoured to avail myſelf of what has been written on it by my predeceſſors with the appearance of reaſon, as well as of the converſation of learned and ingenious men now living, together with ſuch experiments and obſervations as have occurred to me, ſo as to furniſh the means of a more extenſive knowledge of the ſubject.

THE

THE

NATURAL HISTORY

OF THE

TEA-TREE.

PART THE FIRST.

SECT. I

CLASS XIII. ORDER III.

POLYANDRIA TRIGYNIA *(a)*.

K. Calyx, Fig. 1. 2. 3. 10.	Perianthium quinquepartitum, *minimum*, *planum*, ſegmentis *rotundis*, *obtuſis*, *perſiſtentibus*. (Fig. 1. K.)	K. The Calyx, Fig. 1. 2. 3. 10.	A Perianthium quinquepartite, very ſmall, flat, the ſegments round, obtuſe, permanent. (Fig. 1. K.)

(a) Linnæus places the Tea under the order of Monogynia, but the plant in the duke of Northumberland's garden at Sion, flowered laſt October, which has been the means of rectifying this miſtake of the learned profeſſor. See Amœn. Acad. Vol. VII.

C. Corolla, F. 1. 3. 4. 5. 6. 7. 8.	Petala *ſex*;	C. The Corolla, F. 1. 3. 4. 5. 7. 8.	The Petals ſix *(b)*,
	ſubrotunda,		ſubrotund, or roundiſh.
	concava :		concave :
	duo exteriora (F. 4. 7. C. C.)		two exterior, (F. 4. 7. C. C.)
	minora,		leſs,
	inæqualia,		unequal,
	Florem nondum expanſum circumdantia: (F. 3. C.)		incloſing the flower before it is fully blown: (F. 3. C.)
	quatuor interiora, (F. 6. C. C. C. C. & F. 5.)		four interior, (F. 6. C. C. C. C. and F. 5.)
	magna,		large,
	æqualia,		equal,
	antequam decidunt, recurvata. (F. 8. C. C.)		before they fall off, recurvate. (F. 8. C. C.)

(*b*) Among ſome hundred ſpecimens of dried Tea-flowers that I have examined, ſcarcely one in twenty was perfect. Some had three petals only, ſome nine, and others the ſeveral intermediate numbers. The flowers which ſeemed complete in their number, conſiſted of ſix large petals, and externally three leſſer ones of the ſame form. But the flowers which bloſſomed on the Tea-plant belonging to the duke of Northumberland, conſiſted in general of ſix petals, from which this deſcription is taken. One of the flowers indeed, appeared to have eight petals; however, the number in the flowers of moſt plants vary conſiderably, which may account for the miſtake of the indefatigable Dr. Hill, and profeſſor Linnæus, (who deſcribed this plant on Dr. Hill's authority) who make the green and bohea Tea two diſtinct ſpecies, giving nine petals to the former, and ſix to bohea. See Amœn. Acad. Vol. VII. p. 248. Hill. Exot. t. 22. Kæmpfer. Amœn. Exot. p. 607. Breyn. Exot. Plant. Cent. 1. p. 111.

Stamina,

STAMINA, F. 6. 9. 10. 11.
- f. FILAMENTA *numeroſa*, (ducenta circiter.) (f. a. F. 6. 9.)
- *filiformia*,
- *corolla breviora*.
- a. ANTHERÆ cordatæ, biloculares. (F. 10. 11. * Lente aucta.)

The STAMENS, F. 6. 9. 10. 11.
- f. The FILAMENTS numerous *(c)*, (f. a. Fig. 6. 9.) (about 200).
- filiform,
- ſhorter than the Corolla.
- a. The ANTHERAS cordate, bilocular *(d)*. (F. 10. 11. * magnified.)

PISTILLUM, F. 1. 10. 12. * Lente auctum.
- g. GERMEN *globoſo-trigonum*. (F. 1 10. 12.)
- s. STYLI tres, ad baſin coaliti, (F. 12.)
- *ſubulati*,
- recurvati,
- *longitudine ſtaminum*,
- inter ſtamina conferta coarctati & velut in unum

The PISTIL-LUM, F. 1. 10. 12. * magnified
- g. The GERMEN three globular bodies joined in a triangular form. (F. 1. 10. 12.)
- s. The STYLES three, connected at their baſe, (F. 12.)
- ſubulate,
- recurvate,
- of the length of the ſtamens,
- preſſed together, and as if united in one by the thickſet

(c) In a flower I received from that accurate naturaliſt, J. Ellis, F. R. S. &c. I counted upwards of 280 filaments.

(d) Kæmpfer deſcribes the Antheras as being ſingle.

conſolidati,

Pistillum, F. 1. 10. 12. * Lente auctum.	conſolidati, (F. 6. 9. 10.) Petalis autem Staminibuſque delapſis, a ſe mutuo recedentes, divaricantes, & longitudine aucta, marceſcentes. (F. 1. 12.) t. Stigmata *ſimplicia.* (F. 1. 9. 10. 12.)	The Pistillum, F. 1. 10. 12. * magnified	ſurrounding ſtamens *(e)*, (F. 6. 9. 10.) but, after the petals and ſtamens are fallen off, they part from each other, ſpread open, increaſe in length, and wither on the Germen. (F. 1. 12.) t. The Stigmas ſimple. (F. 1. 9. 10. 12.)
P. Pericarpium, F. 1. 13. 14.	Capsula *ex tribus globis coalita*, (F. 13.) trilocularis, (F. 14.) apice trifariam dehiſcens. (F. 13.)	P. The Pericarpium, F. 1. 13. 14.	A Capsule in the form of three globular bodies united, (F. 13.) trilocular, (F. 14.) gaping at the top in three directions. (F. 13.)
S. Semina, F. 14.	*ſolitaria,* *globoſa,* *introrſum angulata.*	S. The Seeds, F. 14.	ſingle, globoſe, angulate on the inward ſide.

(e) This has occaſioned the miſtake of profeſſor Linnæus, in placing this plant under the order of Monogynia. The deception is very natural from examining dried ſpecimens only. See pag. 1. note (*a*).

T. Trun-

T. TRUNCUS, F. I.	ramoſus,	T. The TRUNK (*f*), F. I.	ramoſe,
	lignoſus,		ligneous,
	teres ·		round :
	ramis alternis,		the branches alternate,
	vagis,		vague, *or placed in no regular order*,
	rigidiuſculis,		ſtiffiſh,
	cineraſcentibus,		inclining to an aſh color,
	prope apicem rufeſcentibus.		towards the top reddiſh.
	Florum pedunculi axillares, (F. I. p.)		The peduncles axillary, (F. I. p.)
	alterni,		alternate,
	ſolitarii,		ſingle,
	curvati,		curved,
	uniflori,		uniflorous,

(*f*) Authors widely differ reſpecting the ſize of this tree. Le Compte ſays, it grows of various ſizes from two feet to two hundred, and ſometimes ſo thick that two men can ſcarcely graſp the trunk in their arms: though he afterwards obſerves, that the Tea-trees he ſaw in the province of Fokien, did not exceed five or ſix feet. Journey through the empire of China. London, 1697, 8vo. p. 228. Du Halde quotes a Chineſe author, who deſcribes the height of different Tea-trees, from one to thirty feet. Deſcription gènerale hiſtorique, chronologique, politique, et phyſique de la Chine, Paris, 1755. Fol. 4 Tom. Hiſtory of China, London, 1736. 8vo. Vol. IV. page 22. See alſo Le Spectacle de la Nature, par l'Abbè Pluche, Tom I. p. 486. Edit. 1732. á Paris. Concorde de la geographie, 1754. á Paris, ouvrage poſthume.

But Kæmpfer, who is chiefly to be depended upon, confines the full growth to about a man's heighth. Amœn. Exot. Lemgov. 1712, 4to. page 605. Probably this may be a juſt medium, for Oſbeck ſays, that he ſaw Tea-ſhrubs in flower-pots, not above an ell high. Voyage to China, Vol. I. p. 247. See alſo Eckeberg's account of the Chineſe huſbandry, Vol. II. p. 303.

T. Truncus, F 1.	incraſſati, (F. 1. 2. 7.) ſtipulati: ſtipula ſolitaria, ſubulata, } (F. 1. 2. erecta. } 7. 9. d.)	T. The Trunk, F. 1.	incraſſate, (F. 1. 2. 7.) *(g)* ſtipulate: the ſtipula ſingle, ſubulate, } (F. 1. 2. erect. } 7. 9. d.)
F. Folia, F 1. 15. 16. 17.	alterna, elliptica, obtuſe ſerrata, marginibus inter dentes recurvatis; apice marginata, (F. 15. e.) } * Lente baſi integerrima, (F. 16. 17.) } aucta. glabra, nitida, bullata, ſubtus venoſa,	F. The Leaves, F. 1. 15. 16. 17.	alternate, elliptical, obtuſely ſerrate, with the edges between the teeth recurvate, with the apex emarginate, (F. 15. e.) *(h)* } * mag- at the baſe very entire, (F. 16. 17.) } nified. the ſurface ſmooth, gloſſy, bullate *(i)*, venoſe on the under ſide,

(g) When the peduncles encreaſe in thickneſs towards their extremities, being thinner at their inſertions into the trunk.

(h) No author has hitherto remarked this obvious circumſtance, even Kæmpfer himſelf ſays, that the leaves terminate in a ſharp point. Amœn. Exot. p. 611.

(i) When the upper ſurface of the leaf riſes in ſeveral places in roundiſh ſwellings, hollow underneath.

conſiſtentia,

Latin		English	
F. Folia, F. 1. 15. 16. 17.	consistentia, petiolata:	F. The Leaves, F. 1. 15 16. 17.	of a firm texture, petiolate:
	Petiolis brevissimis, (F. 1. 16. 17. b.)		The Petioles very short, (F. 1. 16. 17. b.)
	subtus teretibus, gibbis, (F. 16. b. * Lente auctis.)		round on the under side, gibbous, *or* *bunching* out, (F. 16. b. * magnified.)
	supra plano-canaliculatis. (F. 17. b.* Lente auctis.)		on the upper-side, flattish, and slightly channelled. (F. 17. b. * magnified.)
	Nomina trivialia Thea bohea & viridis.		The common names bohea and green Teas *(k)*.

There is only one species of this plant; the difference of green and bohea Tea depending upon the nature of the soil, the culture, and manner of drying the leaves. It has even been observed, that a green Tea tree, planted in the bohea country, will produce bohea Tea, and so the contrary *(l)*.

(k) Whether the word Tea, is borrowed from the Japanese *Tsjaa*, or the Chinese *Theb*, is not of much importance. By this name, with vety little difference in pronunciation, the plant here treated of is well known in most parts of the world.

(l) I have examined several hundred flowers, both from the bohea and green Tea countries, and their botanical characters have always appeared uniform. See Directions for bringing over seeds and plants from distant countries, by J. Ellis, Esq.

SECT.

SECT. II.

SYNONIMA.

Many authors have at different times treated upon this ſubject; ſome who never ſaw the Tea-tree, as well as others who had (m). I ſhall firſt enumerate thoſe which are mentioned in the ſpecies plantarum of Linnæus (n).

Thea, Hortus Cliffort. 204. Mat. Med. 264. Hill. Exot. t. 22.

Thee, Kæmpfer. Japan. 605. t. 606.

Thee frutex. Barthol. Act. 4. p. 1. t. 1. Bont. Jav. 87. to 88.

Thee Sinenſium. Breyn. Cent. 111. t. 112. icon. 17. t. 3. Bocc. Muſ. 114. t. 94.

Chaa. Bauh. pin. 147.

Evonymo affinis arbor orientalis nucifera, flore roſeo. Pluk. Alm. 139. t. 88. fig. 6.

In the Acta Haffnienſia, we meet with the firſt figure of this tree; but as it was taken from a dried ſpecimen, it does not illuſtrate the ſubject very well. Bontius publiſhed another, and though drawn in India, where he might have ſeen the plant, it does not much ſurpaſs the preceding. The figure given by Plukenet is better than either of the former; and after his, Breynius publiſhed one ſtill better: But the moſt accurate figure, as well as the beſt deſcription, is given by Kæmpfer (o), and even this figure has ſo many faults,

(m) See Jac. Breynii. Exotic. Cent. 1. p. 114, 115.

(n) Vol. I. p. 734.

(o) Amœnit. Exotic. p. 618, et ſeq. See alſo his hiſtory of Japan by Scheuchzer. Lond. 2 Vol. Fol. App. p. 3. Geoffr. Mat. Med. Vol. II. p. 276.

that

that it may be doubted, whether it were not drawn from an imperfect dried ſpecimen, or ſome mutilated plant, which had paſſed through the fingers of the expert Chineſe (*p*).

SECT. III.

Beſides the authors quoted above, ſeveral others have given ſome account of this exotic ever-green, the principal of which are here added, that the reader who requires further information may conſult the ſame (*q*).

Johann. Petr Maffeus rerum Indicarum libro VI. pag. 108, & lib. XII. p. 242. Ludov. Almeyd. in eodem opere lib. IV. ſelect. epiſt.

Petr. Jarric. Tom. II. lib. II. cap. XVII.

Matth. Ric. de Chriſtian. exped. apud Sinas, lib. I. cap. VII.

Alois Frois, in Relat. Japonicâ.

Nicol. Trigaut. de Regno Chinæ, cap. III. p. 34.

Linſcot. de Inſula Japonicâ, cap. XXVI. pag. 35.

Bernhard. Varen. in deſcriptione Regni Japoniæ, cap. XXIII. pag. 161.

(*p*) Oſbeck in his voyage to China, ſpeaking of the Camellia, ſays, " I bought one of a blind man in the ſtreet, which had fine double white and red flowers. But by further obſerving it in my room, I found that the flowers were taken from another; and one calyx was ſo neatly fixed in the other with nails of Bamboo, that I ſhould ſcarce have found it out, if the flowers had not begun to wither. The tree itſelf had only buds, but no open flowers. I learned from this inſtance, that whoever will deal with the Chineſe, muſt make uſe of his utmoſt circumſpection, and even then muſt run the riſk of being cheated." Vol. VII. p. 17.

(*q*) Vid. Jac. Breynii Gedanenſis Exoticarum aliarumque minus cognitarum Plantarum Cent. I. 1678, Fol. pag. 114.

Joh. Bauhin. Hiſtor. Univerſ. Plantarum, 1597. Tom. III. lib. XXVII. cap. I. pag. 5. b.

Alex. Rhod. *Sommaire des divers voyages et Miſſions Apoſtoliques du R. P. Alexandre de Rhodes de la compagnie de Jeſus á la Chine et autres Royaumes de l'orient, avec ſon retour de la Chine, á Rome; depuis l'anneé* 1618, *juſques á l'année* 1653, *p.* 25.

Les Lettres curieuſes et edifiantes des Jeſuits.

Nicol. Tulpii. Obſervat. Medic. lib. IV. cap. LX. p. 380. Leidæ, 1641. 8vo.

Adam. Olearii *Perſianiſche Reiſe-Beſchreibung*, lib. V. cap. XVII. p. 599. Fol. 1656. Hamburg, 1696. Amſtelod. 1666. 4to.

Johan. Albert. *von Mandelſlo, Morgenlandiſche Reiſe-Beſchreibung*, lib. I. cap. XI. pag. 39. Edit. 1656.

Olai Wormii, Muſ. lib. II. cap. XIV. pag. 165.

Dionyſii Joncquet, Stirpium aliquot paulò obſcurius officinis, Arabibus aliiſque denominatarum, per Caſp. Bauhin. explicat. p. 25. Ed. 1612.

Simon Pauli Comment. de Abuſu Tobaci et Herbæ Thee. Straſburgh. 1665. Lond. 1746.

Simon Pauli Quadripartitum Botanicum, Claſſe ſecundâ, pag. 44. Ibidemque claſſe tertiâ, pag. 493.

Wilhelm. Leyl. epiſtol. apud Simon Pauli in Comment. de Abuſu Tabaci, &c. p. 15. b.

Joann. Nieuzofs, *Gezantſchap an den Keizer van* China, pag. 122. a.

Eraſmi Franciſſ. *Oſt-und Weſt-Indiſcher wie auch Sineſiſcher Luſt-und Stats-Garten*, p. 291.

Oliv. Dappers *Beſchryvinge des Keizerryts van Taiſing of Sina*, Amſtel. 1680. Fol. p. 226.

Athanas.

Athanas. Kircher, Chin. Illuſtrat. Ed. 1658.

Pechlin Theophilus bibaculus, Franckfort, 1684.

Le Compte's journey through the empire of China. London, 1697. 8vo. pag. 228.

Joh. Ludov. Apinus, Obſ. 70. Decur. 3. Miſcell. Curios. 1697. Andr. Cleyerus, Dec. 2. An. 4ti. pag. 7. Dan. Crugerus, Dec. 2. Ann. 4ti. p. 141. Riedlinus, Lin. Med. Ann. 4ti. Dom. Ambroſ. Stegmann, de Decoct. Theæ. Vol. V. p. 36.

Chamberlain's treatiſe of Coffee, Tea, and Chocolate. Lond. 1685. 12mo. p. 46.

Sir Thomas Pope Blount's Natural Hiſtory, 8vo. London, 1693.

Philoſophical Tranſactions, Vol. III. No. 14. London, 1712.

Kæmpfer. Amœnit. Exotic. Lemgov. 4to. 1712. p. 618.

——— hiſtory of Japan by Scheuchzer. Lond. 2 V. Fol. Append. p. 1, & ſeq.

Labat Nouveau voyage aux Iles de l'Amerique. Paris, 1721.

Short's Diſſertation upon the nature and properties of Tea, &c. 4to. London, 1730.

Maſon on the properties of Tea.

Ancient accounts of India and China, by two Mahommedan Travellers. London, S. Harding, 1732.

L'Abbé Pluche Le Spectacle de la Nature, á Paris, 1732.

Du Halde Deſcription gènèral Hiſtorique, Chronologique, Politique et Phyſique de la Chine, Paris, Fol. 4 Vol. Hiſtory of Japan, London, 4 Vol. 8vo. 1735.

Caſp. Neumann, *Vom Thee, Coffee, Bier, und Wein*, Leipſ. 1735.

Chambers' Encyclopædia, Tom 2.

Aſtley's Collection of Voyages, 4 Vol. 4to. London, 1746.

Concorde de la Geographie, á Paris, ouvrage poſthume, 1754.

The good and bad effects of Tea conſidered, Anonymous, London, 8vo. 758.

Linnæi Amœnit. Acad. Vol. VII. p. 241.

Neumann's Chemiſtry, by Lewis, 4to. 1759, pag. 373.

Hanway's Journal of eight days journey, 2 Vol. London, pag. 21. Vol. II.

Hart's Eſſays on Huſbandry, pag. 166.

Percival's Experim. and Medical Eſſays, 8vo. pag. 119.

Oſbeck's voyage into China, by Forſter, London, 2 Vol. 8vo.

Young's Farmer's Letters, Vol. I. p. 299, & 202.

Tiſſot on diſeaſes incidental to literary and ſedentary perſons, by Kirkpatrick, London, 1769. 12mo. pag. 145.

Bomaire Dictionaire d'Hiſtoire naturelle, 8vo. á Paris, 1769.

Milne's Botanical Dictionary, 8vo. London, 1770.

SECT. IV.

ORIGIN OF TEA.

As China and Japan *(r)* are the only countries known to us, where the Tea ſhrub is cultivated, we may reaſonably conclude, that it is indigenous to one of them, if not to both. What motive firſt led the natives to uſe an infuſion of Tea in the preſent manner is uncertain; but probably in order to corr ct the water, which is ſaid to be brackiſh and ill taſted in many parts of thoſe countries *(s)* Of the good effects of

(r) Some authors add Siam alſo.

(s) Le Compte's journey through the empire of China, p. 112.

Tea in ſuch Caſes, we have a remarkable proof in Kalm's journey through North America, which his tranſlator gives us in the following words:

" Tea is differently eſteemed by different people, and I think we *would* be as well, and our purſes much better, if we were without tea and coffee. However, I muſt be impartial, and mention in praiſe of Tea, that if it be uſeful, it muſt certainly be ſo in ſummer, on ſuch journies as mine, through a deſart country, where one cannot carry wine or other liquors, and where the water is generally unfit for uſe, as being full of inſects. In ſuch caſes it is very pleaſant when boiled, and Tea is drank with it; and I cannot ſufficiently deſcribe the fine taſte it has in ſuch circumſtances. It relieves a weary traveller more than can be imagined, as I have myſelf experienced, together with a great many others, who have travelled through the deſart foreſts of America: on ſuch journies Tea is found to be almoſt as neceſſary as victuals *(t)*."

This article was firſt introduced into Europe by the Dutch East India company, very early in the laſt century; and a quantity of it was brought over from Holland about the year 1666 *(u)*, by lord Arlington and lord Oſſory. It ſoon became known amongſt people of faſhion, and its uſe by degrees ſince that period is become univerſal.

It is indeed certain, that before this time, drinking Tea even in public coffee-houſes was not uncommon; for in 1660,

(t) Kalm's travels into North America, Vol. II. p. 314. The following note is added by the ingenious Engliſh tranſlator:

" On my travels through the deſart plains, beyond the river Volga, I have had ſeveral opportunities of making the ſame obſervations on Tea, and every traveller in the ſame circumſtances, will readily allow them to be very juſt." Forſter ibid.

(u) Hanway's Journal of eight days journey, Vol. II. p. 21. The ſame author obſerves, that Tea ſold at this time for ſixty ſhillings a pound.

a duty

a duty of 8d. per gallon was laid on the liquor, made and ſold in all coffee-houſes *(x)*.

So early as 1679, Cornelius Bontekoe, a Dutch phyſician, publiſhed a treatiſe in Dutch, on tea, coffee, and chocolate. In this he ſhews himſelf a very zealous advocate for Tea, and denies the poſſibility of its injuring the ſtomach, although taken to the greateſt exceſs, as far as one or two hundred cups in a day. Whether or no political intereſt might influence Dr. Bontekoe, is uncertain; but as he was firſt phyſician to the Elector of Brandenburgh, and probably of conſiderable eminence and character, his eulogium might tend greatly to promote its uſe: however we find its importation and conſumption were daily augmented, and before the concluſion of the laſt century, it became generally known among the common people in England.

It is foreign to my ſubject, or it would perhaps afford to a ſpeculative mind, no inconſiderable ſatisfaction, to trace the conſumption from its firſt entrance at the Cuſtom-houſe, to the preſent amazing imports. I have been told, that at leaſt three millions of pounds are annually allowed for home conſumption *(y)*; and that the Eaſt-India company have generally in their warehouſes a ſupply for three years.

It is probable that the Dutch, as they traded conſiderably to Japan about the time Tea was introduced into Europe, firſt brought this article from thence. But now China is the general mart, and the province Fokien *(z)*, is the principal

(x) Short's introductory preface to the natural hiſtory of Tea, p. 13.

(y) The quantity of Tea annually ſmuggled into this kingdom is almoſt incredible; which is not included in the above calculation.

(z) In this province, this ſhrub is called Thee or Te; and as the Europeans firſt landed here, that dialect has been preſerved. Le Compte's journey through the empire of China, p. 227. Du Halde's hiſtory of China, Vol. IV. p. 21.

country,

country, that ſupplies both the empire and Europe with this commodity.

S E C T. V.

SOIL AND CULTURE.

We are principally indebted to Kæmpfer, for any accounts that may be relied on, in reſpect to the method of cultivation; and his deſcription was drawn up in Japan. We ſhall give what he ſays upon this ſubject, and then ſtate the accounts we have been able to collect of the Chineſe method.

Kæmpfer tells us, that no particular gardens or fields are allotted for this plant, but that it is cultivated round the borders of the fields, without any regard to the ſoil. Any number of the ſeeds, as they are contained in their ſeed-veſſels, not uſually leſs than ſix, or exceeding twelve or fifteen, are promiſcuouſly put into one hole, made four or five inches deep in the ground, at certain diſtances from each other. The ſeeds contain a large proportion of oil, which is ſoon liable to turn rancid; hence ſcarce a fifth part of them germinate, and this makes it neceſſary to plant ſo many together.

The ſeeds vegetate without any other care; but the more induſtrious annually remove the weeds, and manure the land. The leaves which ſucceed are not fit to be plucked before the third year's growth, at which period they are plentiful, and at their prime.

In about ſeven years the ſhrub riſes to a man's height; but as it then bears few leaves, and grows ſlowly, it is cut down to the ſtem, which occaſions ſuch an exuberance of freſh

fresh shoots and leaves the succeeding summer, as abundantly compensates the owners for their former loss and trouble. Some defer cutting them till they are of ten ears growth.

So far as can be gathered from authors and travellers of credit, this shrub is cultivated and prepared in China, in a similar manner to what is practised in Japan; but as the Chinese export considerable quantities of Tea, they plant whole fields with it, to supply foreign markets, as well as for home consumption.

The Tea-tree delights particularly in vallies, or on the declivities of hills, and upon the banks of rivers, where it enjoys a southern exposure to the sun; though it endures considerable variations of heat and cold, as it flourishes in the northern clime of Pekin, as well as about Canton *(a)*, the former of which is in the same latitude with Rome; and from meteorological observations it appears, that the degree of cold about Pekin is as severe in winter, as in some of the northern parts of Europe *(b)*.

(a) The best Tea grows in a mild temperate climate: The country about Nankin producing better Tea than either Pekin or Canton, betwixt which places it is situated. It has been asserted, that no Tea-plants have yet died in England through excess of cold; but an example of the contrary I know has happened. The plant in the Princess Dowager's garden at Kew, flourishes under glass windows, with the natural heat of the sun, as well as those at Mile-end, in the possession of the indefatigable J. Gordon. Two of the Tea-plants belonging to Dr. Fothergill thrive in his garden at Upton, exposed to the open air in summer.

(b) Du Halde and other authors have observed, that the degree of cold in some parts of China is very severe in winter. In the inland parts of North America, and on extensive continents, the degrees of heat and cold are found to be much more violent than in islands or places bordering on the sea in the same latitude, as the air that blows over the sea is liable to less variation in these respects, than that which blows over large tracts of land, because the sea, large lakes, &c. continue near one temperature through different seasons.

SECT.

SECT. VI.

GATHERING THE LEAVES.

At the proper ſeaſons for gathering the Tea leaves, labourers are hired, who are very quick in plucking them, being accuſtomed to follow this employment as a means of their livelihood. They do not pluck them by handfuls, but carefully one by one; and tedious as this may appear, they are able to collect from four to ten or fifteen pounds each, in one day. The different periods in which the leaves are uſually gathered, are particularly deſcribed by Kæmpfer (*c*).

I. The firſt commences at the middle of the firſt moon, preceding the vernal equinox, which is the firſt month of the Japaneſe year, and falls about the latter end of our February, or beginning of March. The leaves collected at this time are called Ficki Tsjaa, or powdered Tea, becauſe they are pulveriſed and ſipped in hot water (SECT. IX. 1). Theſe tender young leaves are but a few days old when they are plucked; and becauſe of their ſcarcity and price, are diſpoſed of to princes and rich people only; and hence this kind is called imperial Tea.

A ſimilar ſort is alſo called Udſi Tsjaa, and Tacke Sacki Tsjaa, from the particular places where it grows. The peculiar care and nicety obſerved in gathering the Tea leaves in theſe places may deſerve ſome notice here, and we ſhall therefore give ſome account of one of them.

(*c*) Amœnitat. Exotic. p. 618, et ſeq. Hiſtory of Japan. Appendix to Vol. II. p. 6, et ſeq.

Udſi is a ſmall Japaneſe town, bordering on the ſea, and not far diſtant from the city of Miaco. In the diſtrict of this little town, is a pleaſant mountain of the ſame name, which is reckoned to poſſeſs the moſt favorable ſoil and climate for the culture of Tea, on which account it is incloſed with hedges, and likewiſe ſurrounded with a broad ditch for further ſecurity. The trees are planted upon this mountain in ſuch a manner as to form regular rows, with intervening walks. Perſons are appointed to ſuperintend the place, and preſerve the leaves from injury or dirt. The labourers who are to gather them, for ſome weeks before they begin, abſtain from every kind of groſs food, or whatever might endanger communicating any ill flavor; they pluck them alſo with the ſame delicacy, having on a thin pair of gloves *(d)*. This ſort of imperial or bloom Tea *(e)*, is afterwards prepared, and then eſcorted by the chief ſurveyor of the works of this mountain, with a ſtrong guard, and a numerous retinue, to the emperor's court, for the uſe of the imperial family.

II. The ſecond gathering is made in the ſecond Japaneſe month, about the latter end of March, or beginning of April. Some of the leaves at this period are come to perfection, others not arrived at their full growth; both however are promiſcuouſly gathered, and are afterwards ſorted into different claſſes, according to their age, ſize, and goodneſs; the youngeſt particularly are carefully ſeparated, and are often ſold for the firſt gathering or imperial Tea. The tea col-

(d) The ſame cautions are not uſed previous to collecting other ſorts of Tea.

(e) This cannot be the ſort to which alſo the Dutch give that name, as it is ſold upon the ſpot to the princes of the country, for much more than the common bloom Tea is ſold for in Europe. Kæmpfer. Amœnit. Exotic. p. 617. Hiſtory of Japan. Appendix, p. 9. Neumann's Chemiſtry by Lewis, p. 373.

lected at this time is called Tootsjaa, or Chinese Tea, because it is infused, and drank after the Chinese manner (SECT. IX. 1). It is divided by the Tea dealers and merchants into four kinds, distinguished by as many names.

III. The third and last gathering, is made in the third Japanese month, which falls about our June, when the leaves are very plentiful and full grown. This kind of Tea called Ban Tsjaa, is the coarsest, and is chiefly drank by the lower class of people (SECT. IX. III)

Some confine themselves to two gatherings in the year, their first and second, answering the preceding second and third. Others have only one general gathering (*f*), which they make also at the same time with the preceding third or last gathering: however, the leaves collected at each time, are respectively separated into different sortments.

We have observed (SECT. V.), that the Tea tree frequently grows on the steep declivities of hills and precipices, where it is commonly dangerous, sometimes impracticable to collect the leaves, which are often the finest Tea. The Chinese in some places surmount this difficulty by a singular contrivance. These cliffs are inhabited by a large kind of monkies; these the Tea gatherers irritate by some means; in revenge the monkies break off the branches of the Tea tree, and throw them down in resentment; the branches are gathered up, and the Tea leaves picked off. This method of coming at the Tea in such places, was pointed out to me upon some curious Chinese drawings, representing the whole process of

(*f*) In this case the under leaves, which are harsh and less succulent, are probably left upon the trees. See Eckeberg's Chinese husbandry in Osbeck's voyage, Vol. II. p. 303.

 gathering

gathering and curing Tea; and I have ſince been informed by a very inquiſitive ſenſible commander, who has been long in the Company's ſervice, and frequently at China, that this circumſtance is a well known fact.

The Chineſe collect the Tea at certain ſeaſons (*g*), but whether the ſame as in Japan, we are not ſo well informed, moſt probably, however, the Tea harveſt is nearly at the ſame periods, as the natives have frequent intercourſe, and carry on a conſiderable trade with each other (*h*).

S E C T. VII.

METHOD OF CURING OR PREPARING TEA.

Public buildings or drying houſes are erected for curing Tea, and ſo regulated, that every perſon, who either has not ſuitable conveniences, or wants the requiſite ſkill, may bring his leaves at any time to be dried. Theſe buildings contain from five to ten or twenty ſmall furnaces, about three feet high, each having at the top a large flat iron pan (*i*), either

(*g*) Du Halde's hiſtory of China, Vol. IV. p. 21.

(*h*) Ibid. Vol. II. p. 300. Kæmpfer obſerves in his hiſtory of Japan, that the trade between theſe nations has continued from remoteſt antiquity; formerly the Chineſe had a much more general intercourſe with the Japaneſe than they have at preſent; the affinity in the religion, cuſtoms, books, learned languages, arts and ſciences of the Chineſe with the latter, had procured them a free toleration in Japan. Hiſtory of Japan, Vol. I. p. 374.

(*i*) Some writers mention copper pans, and ſuppoſe, that the green efflorescence which appears on copper, may encreaſe the verdure of green Tea; but from experiments that I made, there does not appear any foundation for this ſuppoſition. See SECT. VIII.

high,

ſquare or round, bent up a little on that ſide which is over the mouth of the furnace, which at once ſecures the operator from the heat of the furnace, and prevents the leaves from falling off.

There is alſo a long low table covered with matts, on which the leaves are laid, and rolled by workmen, who ſit round it. The iron pan being heated to a certain degree by a little fire made in the furnace underneath, a few pounds of the freſh gathered leaves are put upon the pan; the freſh and juicy leaves crack when they touch the pan, and it is the buſineſs of the operator to ſhift them as quick as poſſible with his bare hands, till they grow too hot to be eaſily endured. At this inſtant he takes off the leaves, with a kind of ſhovel, reſembling a fan, and pours them on the matts to the rollers, who taking ſmall quantities at a time, roll them in the palms of their hands in one direction, while others are fanning them, that they may cool the more ſpeedily, and retain their curl the longer.

This proceſs is repeated two or three times, or oftner, before the Tea is put in the ſtores, in order that all the moiſture of the leaves may be thoroughly diſſipated, and their curl more completely preſerved. On every repetition the pan is leſs heated, and the operation performed more ſlowly and cautiouſly *(k)*. The Tea is then ſeparated into the different kinds, and depoſited in the ſtore for domeſtic uſe or exportation.

As the leaves of the Ficki Tea (SECT. VI. and IX. 11), are uſually reduced into a powder before they are drank, they

(k) This ſhould be carefully attended to, in curing the fine green Teas, to preſerve their verdure and periſhable flavor. See SECT. VIII. ad finem.

ſhould

ſhould be roaſted to a greater degree of dryneſs. As ſome of theſe are gathered when very young, tender, and ſmall, they are firſt immerſed in hot water, taken out immediately, and dried without being rolled at all.

Country people cure their leaves in earthen kettles *(l)*, which anſwer every neceſſary purpoſe at leſs trouble and expence, whereby they are enabled to ſell them cheaper.

To complete the preparation, after the Tea has been kept for ſome months, it muſt be taken out of the veſſels, in which it had been contained, and dried again over a very gentle fire, that it may be deprived of any humidity which remained, or might ſince have been contracted.

The common Tea is kept in earthern pots with narrow mouths; but the beſt ſort of Tea uſed by the emperor and nobility, is put in porcellane or China veſſels. The Bantsjaa or coarſeſt Tea, is kept by the country people in ſtraw baſkets, made in the ſhape of barrels, which they place under the roofs of their houſes, near the hole that lets out the ſmoke, and imagine that this ſituation does not injure the Tea.

This is the relation we have from Kæmpfer of the method in which the Japaneſe collected and cured their Tea. In the accounts of China, authors have in general treated very ſlightly of the cultivation and preparation of Tea. Le Compte *(m)* indeed obſerves, that to have good Tea, the leaves ſhould be gathered while they are ſmall, tender, and juicy. They begin commonly to gather them in the months of March and April, according as the ſeaſon is forward; they

(l) This is alſo done in China. See Eckeberg's Chineſe huſbandry in Oſbeck's voyage, Vol. II. p. 303.

(m) Journey through the empire of China.

afterwards

afterwards expoſe them to the ſteam of boiling water to ſoften them; and as ſoon as they are penetrated by it, they draw them over copper plates *(n)* kept on the fire, which dries them by degrees, till they grow brown, and roll up of themſelves in that manner we ſee them.

However it is certain, from the Chineſe drawings, which exhibit a faithful picture, though rudely executed, of the whole proceſs from beginning to end, that the Tea tree grows for the moſt part in hilly countries, on their rocky ſummits, and ſteep declivities, inacceſſible in many places; and it would ſeem by the pains the Chineſe are at, in making paths, fixing a kind of ſcaffolds, and even employing the vengeance of monkeys to aſſiſt them, that theſe places afford the fineſt Tea. It appears from theſe drawings, that the trees in general are not much taller than man's height: The gatherers of the leaves are never repreſented but on the ground, they make uſe of hooked ſticks indeed, but theſe ſeem rather intended to draw the branches towards them, when the trees hang over brooks, rivers, or inacceſſible places, than to bend down the tops or upper branches of the trees on plain ground.

They pick the leaves as ſoon as gathered into different ſorts, and cure them nearly in the manner deſcribed to be practiſed by the Japaneſe. They build a range of ſtoves, like thoſe in a chymiſt's laboratory, or great kitchen, where the men work, and curl the leaves in the pans themſelves. It ſeems alſo that they repeat the drying. They dry it likewiſe, after having ſpread it abroad in ſhallow baſkets, in the

(n) Upon this ſubject, See-SECT. VII. and VIII. It may be doubted alſo whether the concluſion of Le Compte's relation is not erroneous, as it is improbable that any leaves ſhould of themſelves take ſo perfect a curl, as that in which Tea is brought into Europe.

ſun;

ſun; and by the means of ſieves, ſeparate the larger from the ſmaller leaves, and theſe again from the duſt.

The Chineſe put the finer kinds of Tea into conic veſſels, like ſugar loaves, made of tutenaque, tin, or lead, covered with neat matting of bamboo; or in ſquare wooden boxes lined with thin lead, dry leaves and paper, in which manner it is exported to foreign countries. The common Tea is put into baſkets, out of which it is emptied, and packed up in boxes or cheſts as ſoon as it is ſold to the Europeans *(o)*.

One thing ſhould be mentioned to their credit; when their harveſt of Tea is finiſhed, each family fails not to teſtify their gratitude to the Giver.

SECT. VIII.

VARIETIES OF TEA

It has been already obſerved (SECT. VI.) that many different ſortments of Tea are made during the times of collecting the leaves, and theſe are multiplied according to the goodneſs of their preparation, by which the varieties of Tea may be conſiderably augmented *(p)*. The diſtinctions with us are much more limited, being generally confined to three principal kinds of green, and five of bohea.

(o) There are ſeveral diſguſting circumſtances attending the preparation of Tea. Oſbeck ſays, the Chineſe ſervants tread the Tea into the cheſts with their naked feet. Voyage to China, Vol. I. p. 252.

(p) Du Halde's hiſtory of China, Vol. IV. p. 21. Oſbeck's voyage to China. Vol. I. p. 246, et ſeq.

I. Thoſe

I. Thoſe of the former are,

I. Bing, imperial, or bloom Tea, with a large looſe leaf, of a light green color, and faint delicate ſmell.

II. Hy-tiann, or hi-kiong, known to us by the name of hyſon Tea, ſo called after an Eaſt-India merchant of that name, who firſt imported it into Europe. The leaves are cloſely curled and ſmall, of a green color, verging towards blue *(q)*.

III. Singlo, or ſanglo, which name it receives, like many other Teas, from the place where it is cultivated.

II. The bohea Teas.

I. Soochuen, or ſutchong, by the Chineſe called ſaatyang, or ſu-tyann. It imparts a yellowiſh green color, by infuſion *(r)*.

II. Camho, or ſoumlo, called after the name of the place where it is gathered; a fragrant Tea with a violet ſmell. Its infuſion is pale.

III. Congo, or bong-fo. This has a larger leaf than the following, and the infuſion is a little deeper colored. It reſembles the bohea in the color of the leaf *(ſ)*.

(q) The Chineſe have another kind of hyſon Tea, which they call hyſon-utchin, with narrow ſhort leaves. Another ſort of green Tea they name go-bé, the leaves of which are narrow and long.

(r) Padre ſutchong has a finer taſte and ſmell than the common ſutchong. The leaves are large and yellowiſh, not rolled up, but expanded, and packed up in papers of half a pound each. It is generally conveyed by caravans into Ruſſia. Without much care it will be injured at ſea. This Tea is rarely to be met with in England.

(ſ) There is a ſort of Tea called lin-kiſam, with narrow rough leaves. It is ſeldom uſed alone, but mixed with other kinds. By adding it to congo, the Chineſe ſometimes make a kind of pekoe Tea. Oſbeck's voyage to China, Vol. I. p. 249.

IV. Pecko, or pekoe, by the Chinese called back-ho, or pack-ho. It is known by having the appearance of small white flowers intermixed with it.

V. Common bohea, called moji by the Chinese, consists of leaves of one color (*t*).

III. There has also been imported a sort of Tea, of a different form from any of the preceding, made up into cakes or balls of different sizes.

I. The largest kind of this cake Tea, that I have seen, weighs about two ounces; the infusion and taste resemble those of good bohea Tea.

II. Another sort, which is a kind of green Tea, is called tio te': it is rolled up in a round shape, about the size of peas.

III. The smallest kind done in this form is called gun powder Tea.

The Chinese likewise prepare an extract from Tea, which they exhibit as a medicine dissolved in a large quantity of water, and ascribe to it many powerful effects in fevers, and other disorders, when they wish to procure a plentiful sweat. This extract is sometimes formed into small cakes, not much broader than a sixpence, sometimes into rolls of a considerable size.

That there is only one species of Tea tree, has already been mentioned (SECT. I.) from which all the varieties of Tea are

(*t*) The best bohea Tea is named by the Chinese tao-kyonn. An inferior kind is called An-kai, from a place of that name. In the district of Honam near Canton, the Tea is very coarse, the leaves yellow or brownish, and the taste the least agreeable of any. By the Chinese it is named Honam té, or Kuli té.

procured.

procured. Kæmpfer, who is of this opinion, attributes the difference of Teas, to the ſoil and culture of the plant, age of the leaves when gathered, and curation (u). Theſe circumſtances will ſeverally have more or leſs influence; though whether they account for all the varieties obſervable in Tea may be doubted.

I infuſed all the ſorts of green and bohea Teas I could procure, and expanded the different leaves on paper, to compare their ſize and texture, and thereby to diſcover their age; I found the leaves of green Tea as large as thoſe of bohea, and nearly as fibrous; which would lead one to ſuſpect that the difference does not ſo much depend upon the age, as upon the other circumſtances.

We know that in Europe, the ſoil, culture, and expoſure have great influence on all kinds of vegetables: the difference is often evident in the ſame province, and even in the ſame diſtrict; but in Japan, and particularly through the continent of China, it muſt be much more conſiderable, where the air is in ſome parts very cold, in others moderate, or warm almoſt to an extreme. I am perſuaded that the method of preparation muſt alſo have no little influence. I have dried the leaves of ſome European plants, in the manner deſcribed (Sect. VI) which ſo much reſembled the foreign Tea, that the infuſion made from them has been ſeen and drunk with- ſuſpicion. In theſe preparations which I made, ſome of the leaves retained a perfect curl, and a fine verdure like the beſt green Tea; and others cured at the ſame time were more like the bohea (x).

(u) This renders what has been obſerved at the concluſion of Sect. I. more probable.

(x) A certain moderate degree of heat preſerved the verdure and flavor better than a haſty exſiccation. In the firſt caſe, it is neceſſary to repeat the roaſting oftener.

 I would

I would not however lay too much ſtreſs upon the reſult of a few trials, nor endeavour to preclude further enquiries about a ſubject, which at ſome future period may prove of more immediate concern to this nation.

We might ſtill try to diſcover whether ſome art is not uſed with Tea before its exportation from China, to produce the difference of color (*y*), and flavor (*z*) peculiar to different ſorts. An intelligent friend of mine informs me, that in a ſet of Chineſe drawings in his poſſeſſion, repreſenting the whole proceſs of making Tea, there are in one ſheet the figures of ſeveral perſons apparently ſeparating the different kinds of Tea, and drying it in the ſun, with ſeveral baſkets ſtanding near them filled with a very white ſubſtance, and in conſiderable quantity. To what uſe this may be applied is uncertain, as well as what the ſubſtance is; yet there is no doubt, he thinks, but it is uſed in the manufacturing of Tea, as the Chineſe ſeldom bring any thing into their pieces but ſuch as relate in ſome reſpect to the buſineſs before them.

As green Tea is by ſome ſuſpected to have been cured on copper, they have attributed the verdure to the effloreſcence from that metal (SECT. VII.), but if there were any foundation for this ſuppoſition, the volatile alkali, mixed with an

(*y*) Infuſions of fine bohea Teas, do not differ a great deal in color from thoſe of green.

(*z*) I am informed by intelligent perſons, who have reſided ſome time at Canton, that the Tea about that city, affords very little ſmell whilſt growing. The ſame is obſerved of the Tea plants in England; and alſo of the dried ſpecimens from China. We are not hence to conclude that art alone conveys to Teas when cured the ſmell peculiar to each kind, for our vegetables, graſſes for inſtance, have little or no ſmell till dried, and made into hay.

infuſion of ſuch Tea, would detect the leaſt portion of copper, by turning the infuſion blue *(a)*.

Others have, with leſs propriety, attributed the verdure to green copperas *(b)*; but this ingredient, which is only ſalt of iron, would immediately turn the leaves black, and the infuſion made from the Tea would be of a deep purple color *(c)*.

Is it not more probable, that ſome green dye, prepared from vegetable ſubſtances, is uſed for the coloring?

SECT. IX.

DRINKING OF TEA.

Neither the Chineſe, nor natives of Japan, ever uſe Tea before it has been kept at leaſt a year; becauſe when freſh it is ſaid to prove narcotic, and diſorder the ſenſes *(d)*. The former pour hot water on the Tea, and draw off the infuſion in the ſame manner as is now introduced from them into Europe; but they drink it ſimply without the addition of

(a) The hundredth part of a grain of copper, diſſolved in a pint of liquor, ſtrikes a ſenſible blue with volatile alkalies. Neumann's chemiſtry, by Lewis, p. 62. The fineſt imperial and bloom Teas ſhewed no ſign of the preſence of this metal by experiment.

(b) See Short on Tea, p. 16. Boerhaave attributed the verdure of green Tea to this ſubſtance.

(c) I remember a diverting incident which happened to a Tea party, who went into the country to ſpend an afternoon together. The water that was boiled for Tea, was brought from a chalybeate ſpring; when this water was poured into the Tea pot on the leaves, it turned immediately like ink, whereby the company were both ſurprized at the phenomenon, and diſappointed of their refreſhment.

(d) Kæmpfer. Amœnit. Exot. p. 625. Hiſtory of Japan, 2 Vol. App. p. 10, 16.

ſugar

ſugar or milk (*e*). The Japaneſe reduce the Tea into a fine powder, by grinding the leaves in a hand-mill, and mix them with hot water into a thin pulp, in which form it is ſipped (*f*), particularly by the nobility and rich people. It is made and ſerved up to company in the following manner: the Tea-table furniture, with the powdered Tea incloſed in a box, are ſet before the company, and the cups are then filled with hot water, and as much of the powder as might lie on the point of a moderate ſized knife, is taken out of the box, put into each cup, and then ſtirred and mixed together with a curious denticulated inſtrument till the liquor foams, in which ſtate it is preſented to the company, and ſipped while warm (*g*). From what Du Halde relates, this method is not peculiar to the Japaneſe, but is alſo uſed in ſome provinces of China (*h*).

The common people, who have a coarſer Tea (Sect. VI. iii.) boil it for ſome time in water, and make uſe of the liquor for common drink. Early in the morning the kettle filled with water, is regularly hung over the fire for this purpoſe, and the Tea is either put into the kettle encloſed in a bag, or by means of a baſket of a proper ſize, preſſed to the bottom of the veſſel, that there may not be any hinderance in drawing off the water. The Bantsjaa Tea, (Sect. VI. iii.) only is uſed in this manner, whoſe virtues being more fixed, would not be ſo fully extracted by infuſion.

And indeed Tea is the common beverage of all the laboring people in China, one ſcarcely ever ſees them repreſented

(*e*) Oſbeck's voyage to China, Vol. I. p. 299.

(*f*) This is called koitsjaa, that is, thick Tea, to diſtinguiſh it from that made by infuſion.

(*g*) An inferior kind of Tea is infuſed, and drank in the Chineſe manner. Sect. VI. ii. and Sect. IX. i.

(*h*) Hiſtory of China, Vol. IV. p. 22.

at

at work of any kind, but the Tea pot and Tea cup are either bringing to them, or ſet by them on the ground. Reapers, threſhers, and all who work out of doors, as well as within, have this attendant (*i*).

To make Tea, and to ſerve it in a genteel and graceful manner, is an accompliſhment, in which people of both ſexes in Japan are inſtructed by maſters, in the ſame manner as Europeans are in dancing, and other branches of a genteel education.

SECT. X.

SUCCEDANEA.

Curioſity and intereſt would mutually induce the Europeans to make the moſt diligent inquiries in order to diſcover the real Tea ſhrub, or a ſubſtitute in ſome other vegetable the moſt reſembling it. Simon Pauli, a phyſician and botaniſt at Copenhagen, was the firſt who pretended to have diſcovered the real Tea plant in Europe. By opening ſome Tea leaves, he found them ſo much like thoſe of the Dutch myrtle (*k*), (Hor. Su. 907). that he obſtinately maintained they were productions of the ſame ſpecies of Tea; though he

(*i*) In public roads, and in all places of much reſort in Japan, and even in the midſt of fields and frequented woods, Tea booths are erected; as moſt travellers drink ſcarcely any thing elſe upon the road. Kæmpfer's hiſtory of Japan, by Scheuchzer. Fol. Vol. II. p. 428.

(*k*) Myrica Gale. Syſtem. Natur. Vol III. p. 651. A plant well known by the name of Gale in the north of England, and indigenous in Brabant, and other northern nations.

was

was aftewards refuted by ſeveral botaniſts in Europe, and by the ſpecimens ſent to him, and to Dr. Mentzel of Berlin, from the Eaſt-Indies by Dr. Cleyer *(l)*.

Father Labat next thought he had diſcovered the real Tea plant in Martinico *(m)*, agreeing, he ſays, in all reſpects with the China ſort. He pretends alſo to have procured Tea ſeeds from the Eaſt Indies, and to have raiſed the plant in America; but from his own account, it appears to be only a ſpecies of Lyſimachia, or what is called Weſt-India Tea *(n)*.

Many other pretended diſcoveries of the oriental Tea-tree have been related; all which have proved erroneous, when properly enquired into. The genus of plant called by Kæmpſer Tsubakki *(o)*, has the neareſt reſemblance. The leaves of ſeveral European plants have been uſed at different times as ſubſtitutes for Tea, either from ſome ſimilarity in the ſhape of the leaves, or in the taſte and flavor; among theſe, two or three ſpecies of Veronica have been particularly re-

(l) Figures of the ſame were publiſhed in the Acta Haffnienſia and German Ephemerides.

(m) Nouveau voyage aux Iles de l'Amerique, Paris, 1721.

(n) This ſhrub I have frequently met with in the Weſt-Indies.

(o) Two ſpecimens of this plant are now in the phyſic garden at Upſal. About the year 1755, they were brought over from China by M. Lagerſtrom, a director of the Swediſh Eaſt-India Company, under the ſuppoſition of being Tea plants, till they appeared in bloſſom, when they proved to be this ſpecies of Tsubakki, called by Linnæus, Camellia. Spec. Plant. p. 982. This celebrated profeſſor ſays, " That the leaves of his Camellia are ſo like the true Tea, that they would deceive the moſt ſkilful botaniſt; the only difference is, that they are a little broader. Amœnit. Academ. Vol. VII. p. 251. See alſo Ellis's directions for bringing over foreign plants, p. 28. A Camellia was laſt ſummer brought from China in good health; the leaves of this ſhrub end in a double obtuſe point, (obtuſely emarginated) like thoſe of the Tea tree, which makes them ſtill more liable to be miſtaken for thoſe of the latter. Kæmpfer obſerves, that the leaves of a ſpecies of Tsubakki are preſerved, and mixed with Tea, to give it a fine flavor. Amœnit. Exotic. p. 858.

commended

commended (*p*), beſides the leaves of ſage, myrtle (*q*), betony, agrimony, wild roſe, and many others (*r*). Whether any of theſe are really more ſalutary or not, we now find, that from the palace to the cottage, every other ſubſtitute has yielded to the genuine Aſiatic Tea.

SECT. XI.

PRESERVING THE SEEDS FOR VEGETATION.

Many attempts to introduce the Tea-tree into Europe, have been unſucceſsfully made, owing to the bad ſtate of the ſeeds when firſt procured, or to want of judgment in preſerving them long enough in a ſtate of vegetation. If this complaint ariſe from the firſt cauſe, future precautions about ſuch ſeeds will be in vain; it is therefore neceſſary to procure freſh, ſound, ripe ſeeds, white, plump and moiſt internally.

Two methods of preſerving the ſeeds have put us in poſſeſſion of a few young plants of the true Tea-tree of China; one is by incloſing the ſeeds in bees wax, after they have been well dried in the ſun; and the other, by putting them, included in their pods, or capſules, into very cloſe caniſters made of tin and tutenague (*s*).

But

(*p*) Veronica officinalis. Flor. Suec. p. 12. Veronica Chamædr. Fl. Suec. p. 18. Pechlin Thephilus bibaculus Franckfort. 1684. Francus de Veronica vel Theezantem.

(*q*) Simon Pauli de abuſu Theæ et Tabaci. Straſburg, 1665. Lond. 1746.

(*r*) See Neumann's chemiſtry, by Lewis, p. 375.

(*s*) See directions for bringing over ſeeds and plants from the Eaſt-Indies, by that great promoter of natural hiſtory, J. Ellis, F. R. S. &c. In which particu-

F lar

But neither of these methods have succeeded generally, notwithstanding the utmost care, both in getting fresh seeds, and in securing them in the most effectual manner. The best method is to sow the ripe seeds in good light earth, at leaving Canton; covering them with wire, to prevent rats and other such vermin coming to them. The boxes should not be exposed to too much air, nor to the spray of the sea if possible. The earth should not be suffered to grow dry and hard, but a little fresh or rain water may be sprinkled now and then; and when the seedling plants appear, they should be kept moist, and out of the burning sun. Most of the plants now in England were procured by these means; and though many of the seedlings will die, yet by this kind of management we may probably succeed in bringing over the most curious vegetable productions of China, and of which they have an amazing treasure, both in respect to use, shew and variety (*t*).

The

lar directions are given, both to choose the proper seeds, and to preserve them in the best manner for vegetation. See also the naturalist's and traveller's companion, containing instructions for discovering and preserving objects of natural history, SECT. III. We may observe here, that the best method of bringing over the parts of flowers intire, is to put them in bottles of spirit of wine, good rum, first runnings or brandy. In the directions, &c. above mentioned, the learned naturalist has not recommended this easy method of preserving the parts of fructification, but in a future edition, I am informed he purposes to do it. Flowers of the Illicium Floridanum, or starry anniseed tree, published in the last vol. of Phil. Trans. (LX.) were sent to him in this manner.

(*t*) Another method has succeeded with some North American seeds, by putting them into a box, not made too close, upon alternate layers of moss, in such a manner as to admit the seeds to vegetate, or shoot their small tendrils into the moss. In the passage, the box may be hung up at the roof of the cabin; and when arrived here, the seeds should be put into pots of mold, with a little of the moss also about them, on which they had lain. This method has procured us seeds in a state fit for vegetation, which had often miscarried under the preceding precautions;

The young Tea plants in the gardens about London thrive very welll in the green houſes in winter, and ſome bear the open air in ſummer. The leaves of many of them are from one to three inches long, not without a fine deep verdure; and the young ſhoots are ſucculent. It is therefore probable, that in a few years many layers may be procured from them, and the number of the plants conſiderably encreaſed thereby.

It may not be improper to obſerve here, that many exotic vegetables, like human conſtitutions, require a certain period before they become naturalized to a change of climate; many plants, which at their firſt introduction would not bear our winters without ſhelter, now endure our hardeſt froſts; the beautiful magnolia, among ſeveral others, is a proof of this obſervation; and we have already taken notice (SECT. V.), that the degree of cold at Pekin, ſometimes exceeds ours. We have therefore reaſon to expect, that the Tea-tree may in a few years be capable of bearing our climate, at length thrive, as if indigenous to this country, and become an article in our exports *(u)*, like the common potatoe, for which we are indebted to America, or Spain *(x)*.

It

tions; and therefore might be tried at leaſt, with Tea and other oriental ſeeds. In order to ſucceed more certainly, ſome of the Tea ſeeds, in whatever manner they may have been preſerved, ſhould be ſown when the veſſel arrives at St. Helena, and alſo after paſſing the tropic of Cancer, near the lat. of 30 deg. north.

(u) The high price of labor in this country, may prove the principal objection to this proſpect. In China proviſions are very cheap. Oſbeck ſays, that a workman who lives upon plucking of Tea leaves, will ſcarce be able to get more than one penny a day, which is ſufficient to maintain him. Voyage to China, Vol. I. p. 298.

(x) Gerard ſays in his herbal, publiſhed Ann. 1597. p. 780. Potatoes grow in India, Barbarie, Spaine, and other hotte regions, of which I planted diuers rootes (that I bought at the exchange in London) in my garden, where they flouriſhed untill winter, at which time they periſhed and rotted." At this date, he adds,

F 2 " they

It is indeed probable that the North American ſummers in the ſame latitude with Pekin, would ſuit this Tree better than ours; for in China, and ſome parts of North America, the heat in ſummer is ſuch, that vegetables make quicker and more early ſhoots, whereby they have time to acquire ſufficient ſtrength and firmneſs, before the winter commences; but in England, the tender ſhoots are puſhed forth late, and winter ſoon after ſucceeding, they often periſh, in a degree of cold much leſs ſevere than at Pekin, or in colder latitudes of North America.

" they were roaſted in the aſhes; ſome when they be ſo roaſted, infuſe them, and ſop them in wine; and others, to give them the greater grace in eating, do boile them with prunes, and ſo eate them. And likewiſe others dreſſe them (being firſt roaſted) with oile, vinegar, and ſalt, euery man according to his own taſte and liking."

THE

THE MEDICAL HISTORY OF TEA.

PART II.

SECT. I.

AS the cuſtom of drinking Tea is become univerſal, every perſon may be conſidered as a judge of it's effects, at leaſt ſo far as it concerns his own health; but as the conſtitutions of mankind, are as various as the individuals, the effects of this infuſion muſt be different alſo, which is the reaſon that ſo many opinions have prevailed upon the ſubject.

Many who have once conceived a prejudice againſt it, ſuffer it to influence their judgment too far, and condemn the cuſtom as univerſally pernicious. Others, who are no leſs biaſſed on the other extreme, would make their own private experience, a ſtandard for the general, and aſcribe the moſt ex-

tenſive

tenſive virtues to this infuſion. This contrariety of opinion has been particularly maintained among phyſicians (*y*), which will ever be the caſe, while mere ſuppoſitions are placed in the room of experiments and facts impartially related.

SECT. II.

There are ſome phyſicians, however, who avoid both extremes; who without commending it or decrying it univerſally, admit it's uſe, without being inſenſible to the injuries received from it. It requires no ſmall ſhare of diſpaſſionate ſagacity to fix the limits of good and harm in the preſent caſe: multitudes of all ages, conſtitutions and complexions, drink it freely, during a long life, without perceiving any ill effects. Others again ſoon experience many inconveniences from drinking any conſiderable quantity of this infuſion.

It is difficult to draw certain concluſions from experiments made on this herb. The parts which ſeem to produce theſe oppoſite effects are very fugitive. We become acquainted chiefly with the groſſer parts by analyſis. I made the following experiments with conſiderable care, but I own they inform us not ſufficiently wherein conſiſts that grateful relaxing ſedative property, that proves to the generality of mankind ſo refreſhing, nor from whence it is, that others feel from this pleaſing beverage many diſagreeable effects. Obſervation muſt inſtruct us in this difficult inveſtigation, more than ſimple experiments on the ſubject itſelf.

(*y*) Compare Joh. Ludov. Hannemane de potu calido in Miſcell. curios. Simon Pauli de abuſu Theæ et Tabaci. Tiſſot on the diſeaſes of literary and ſedentary perſons, &c. with Waldſmick in Diſput. var. argum, &c.

EXPE-

EXPERIMENT I.

I took an equal quantity of an infusion of superfine green Tea, and of common bohea Tea, made equally strong; and also the same quantity of the liquor remaining after distillation (SECT. III. 1.), and of simple water; into each of which, contained in separate vessels, I put two drachms of beef, that had been killed about two days.

The beef which was immersed in the simple water, became putrid in forty eight hours; while the pieces in the two infusions of Tea, and in the liquor remaining after distillation, shewed no signs of putrefaction, till after about seventy hours (*z*).

EXPERIMENT II.

Into strong infusions of every kind of green and bohea Tea that I could procure, I put equal quantities of salt of iron (sal martis), which immediately changed the several infusions into a deep purple color (*a*).

It is evident from these experiments, that both green and bohea Tea possess an antiseptic (EXPER. I.), and astringent power (EXPER. II.), applied to the dead animal fibre.

(*z*) See Percival's Experimental Essays, p. 119, et seq. wherein many ingenious experiments and observations are related.

(*a*) In this experiment, four ounces of infusion were drawn from two drachms of each kind of Tea, and one grain of sal martis added to the respective infusions. See Neumann's chemistry, by Lewis, pag. 377. Short on the nature and properties of Tea, p. 29.

SECT.

SECT. III.

Neverthelefs, as I have often obferved that drinking Tea, particularly the moft highly flavored fine green, proves remarkably relaxing to many perfons of tender and delicate conftitutions, I was induced to profecute my enquiries farther.

1. To this end I diftilled half a pound of the beft and moft fragrant green Tea with fimple water, and drew off an ounce of very odorous and pellucid water, free from oil, and which on trial (SECT. II. EXP. II.), fhewed no figns of aftringency.

2. That part of the liquor which remained after diftillation, was evaporated to the confiftence of an extract; it was flightly odorous, but had a very bitter, ftyptic, or aftringent tafte. The quantity of the extract thus procured weighed about five ounces and a half.

EXPERIMENT III.

a. Into the cavity of the abdomen, and cellular membrane of a frog, about three drachms of the diftilled odorous water (No. 1.) were injected.

In twenty minutes, one hind leg of the frog appeared much affected, and a general lofs of motion and fenfibility fucceeded *(b)*. The affection of the limb continued for

(b) See Smith, Tentamen inaugurale de actione mufculari. Edinb. p. 46.

four

four hours, and the univerſal torpidity remained above nine hours; after this the animal gradually recovered it's former vigor.

b. In like manner ſome of the liquor remaining after the diſtillation of the green Tea, (No. 1.) was injected, but this was not productive of any ſenſible effect.

EXPERIMENT IV.

a. To the iſchiatic nerves laid bare, and to the cavity of the abdomen of a frog, I applied ſome of the diſtilled odorous water (No. 1. and Exp. III, a.). In the ſpace of half an hour, the hindermoſt extremities became altogether paralytic and inſenſible; and in about an hour afterwards the frog died.

b. In like manner I applied the liquor remaining after diſtillation (No. 1. and Exp. III, b.) to another frog, but no ſedative or paralytic effect was obſervable.

c. The extract (No. 2.) diſſolved in water, and applied to the ſame parts under like circumſtances, produced no ſenſible effect.

3. From theſe experiments the ſedative and relaxing effects of Tea, appear greatly to depend upon an odorous fragrant principle, which abounds moſt in green Tea, particularly the highly flavored (*c*). This ſeems further confirmed by the

(*c*) Two drachms of this odorous water were given to a delicate perſon. He was ſoon after affected with a nauſea, ſickneſs, general lowneſs and debility, which continued for ſome hours, which he obſerves uſually ſucceeds when he drinks ſuperfine green Tea.

Smelling forcibly at the ſame has occaſioned ſimilar effects upon ſome delicate people.

practice of the Chinese, who avoid using this plant, till it has been kept at least twelve months, as they find it possesses a soporiferous and intoxicating quality when recent. (PART I. SECT. IX.)

Thus often under trees supinely laid,
Whilst men enjoy the pleasure of the shade,
Whilst those their loving branches seem to spread
To screen the sun, they noxious atoms shed,
From which quick pains arise, and seize the head.
Near Helicon, and round the learned hill
Grow trees, whose blossoms with their odor kill *(d)*.

CREECH.

SECT. IV.

Waving however any attempts to fix with precision, the effects of Tea from these experiments alone, let us endeavour to collect from observation likewise such facts as may enable us to judge what its effects are on the human frame, and from thence draw the clearest inferences we can, how far it is salutary or otherwise.

The long and constant use of Tea, as a part of our diet, makes us forget to enquire whether it is possessed of any me-

(d) Arboribus primum certis gravis umbra tributa est
Usque adeo, capitis faciant ut sæpe dolores,
Siquis eas subter jacuit prostratus in herbis.
Est etiam in magnis Heliconis montibus arbos
Floris odore hominem tetro consueta necare.

LUCRETIUS. B. 6.

dicinal

dicinal properties. We ſhall endeavour to conſider it in both reſpects.

The generality of healthy perſons, find themſelves not apparently affected by the uſe of Tea: it ſeems to them a grateful refreſhment, both fitting them for labor and refreſhing them after it. There are inſtances of perſons who have drank it from their infancy, to old age; have led at the ſame time, active, if not laborious lives; and who never perceived from the conſtant uſe of it any ill effect, nor had any complaint which they could aſcribe to the effects of this liquor.

Where this has been the caſe, the ſubjects were for the moſt part healthy, ſtrong, active, and temperate, both of one ſex and the other. Amongſt the leſs hardy and robuſt, we find complaints, which are aſcribed to Tea, by the parties themſelves. Some complain that after a Tea breakfaſt, they find themſelves rather fluttered; their hands leſs ſteady in writing, or any other employ that requires an exact command. This probably ſoon goes off, and they feel no other effect from it. Others again bear it well in the morning, but from drinking it in the afternoon, find themſelves very eaſily agitated, and affected with a kind of involuntary trembling.

There are many who cannot bear to drink a ſingle diſh of Tea, without being immediately ſick and diſordered at the ſtomach. To ſome it gives great pain about that part, very excruciating, and attended with general tremors. But in general the moſt tender and delicate conſtitutions are moſt affected by the free uſe of Tea; being frequently attacked with pains in the ſtomach and bowels; ſpaſmodic affections; attended with pale limpid urine in large quantities; great agitation of ſpirits, and a proneneſs to be diſconcerted with the leaſt noiſe, hurry or diſturbance.

SECT. V.

There is one circumſtance however that renders it more difficult to inveſtigate the certain effects of Tea; which is, the great unwillingneſs that moſt people ſhew, to giving us a genuine account of their uneaſy ſenſations after the free uſe of it; from a conſciouſneſs that it would be extremely imprudent to continue its uſe, after they are convinced from experience that it is injurious.

That it produces watchfulneſs in ſome conſtitutions, is moſt certain, when drank at evening in conſiderable quantities. Whether warm water would not ſometimes do the ſame, or any other aqueous liquor, is not ſo certain.

That it enlivens, refreſhes, exhilirates, is likewiſe well known. From all which circumſtances it would ſeem, that Tea contains an active penetrating principle, ſpeedily exciting the action of the nerves: in very irritable conſtitutions, to ſuch a degree as to give very uneaſy ſenſations, and bring on ſpaſmodic affections: in leſs irritable conſtitutions, it rather gives pleaſure, and immediate ſatisfaction, though not without occaſionally producing ſome tendency to tremors and agitation bordering upon pain.

The finer the Tea, the more obvious are theſe effects. It is perhaps for this, amongſt other reaſons, that the lower claſſes of people, who can only procure the moſt common, are in general the leaſt ſufferers. I ſay, in general, becauſe even amongſt them, there are many who actually ſuffer much by it: they drink it as long as it yields any taſte, and for the moſt part hot, to add to its flavor; and what the finer kinds

kinds of Tea effect in their superiors, the quantity, and the degree of heat in which it is drank, produce in them.

It ought not however to pass unobserved, that in a multitude of cases, the infusions of our own herbs; sage, for instance, mint, baum, even rosemary, and valerian itself, will now and then produce similar effects, and leave that emptiness, agitation of spirits, flatulence, spasmodic pains, and other symptoms that are met with in people, the most of all others devoted to Tea.

SECT. VI.

That there is something in the finer green Teas, that produces effects peculiar to itself, and not to be equalled by any other substance we know, is I believe admitted by all who have observed, either what passes in themselves, or the accounts that others give of their feelings, after a plentiful use of this liquor. Nor are the finer kinds of bohea Teas exempt from the like influence. They affect the nerves, produce tremblings, and such a state of body for the time, as subjects it to be agitated by the most trifling causes, shutting a door too hastily, the sudden entrance even of a servant, and other the like causes.

I know people of both sexes, who are constantly seized with great uneasiness, anxiety and oppression, as often as they take a single cup of Tea, and who nevertheless, for the sake of company, drink several cups of warm water, mixed with sugar and milk, without the least inconvenience.

A physician whose acquaintance I have long been favored with, and who, with some others, was present when the preceding

preceding experiments were made at the college of Edinburgh, has a remarkable delicacy in feeling the effects of a ſmall quantity of fine Tea. If drank in the forenoon, it affects his ſtomach with an uneaſy ſenſation for ſeveral hours afterwards, and entirely takes away his appetite for food at dinner; though at other times when he takes chocolate for breakfaſt, he generally makes a very hearty meal at noon, and enjoys the moſt perfect health. If he drink a ſingle diſh of tea in the afternoon, it affects him in the ſame manner, and deprives him of ſleep for three or four hours, through the ſucceeding night; yet he can ſocially take a cup of warm water with ſugar and milk, without the leaſt inconvenience.

It may be remarked that opium has nearly the ſame effect upon him as Tea, but in a greater degree; for he informs me, that when he once accidentally took a quantity of the ſolution of opium, it had not the leaſt tendency to induce ſleep, but produced a very diſagreeable uneaſineſs at his ſtomach, approaching to nauſea.

S E C T. VII.

I am informed likewiſe by a phyſician, of long and extenſive practice in the city, that he has known ſeveral inſtances of a ſpitting of blood having been brought on, by breathing in an air loaded with the fine duſt of Tea. It is cuſtomary for thoſe who deal largely in this article, to mix different kinds together, ſo as to ſuit the palates of their cuſtomers in different places. This is generally performed in the back part of their ſhops, ſeveral cheſts perhaps being mixed together at the ſame time. Thoſe who are much employed in this work,

work, are very often ſufferers by it at length; ſome being ſeized with ſudden bleedings from the lungs or from the noſtrils; others attacked with violent coughs, ending in conſumptions.

Theſe circumſtances are chiefly brought in ſight to prove, that beſides a ſedative relaxing power, there exiſts in Tea an active penetrating ſubſtance, which cannot but in many conſtitutions be productive of ſingular effects.

An eminent Tea broker, after having examined in one day, upwards of one hundred cheſts of Tea, by ſmelling at them only, and forcibly, in order to diſtinguiſh their reſpective qualities, was the next day ſeized with a violent giddineſs, head-ach, univerſal ſpaſms, and loſs of ſpeech and memory. By proper aſſiſtance he recovered to a certain degree, but not totally. His ſpeech returned, his memory in ſome degree, his ſtrength never. He continued, with unequal ſteps, gradually loſing ſtrength; a partial paralyſis enſued, then a more general one, and at length he died totally enfeebled and inſenſible. Whether this was owing to the Tea, may perhaps be doubted. Future accidents may poſſibly confirm the ſuſpicions to be juſt or otherwiſe.

S E C T. VIII.

An aſſiſtant to a Tea broker, had frequently for ſome weeks complained of pain and giddineſs of his head, after examining and mixing different kinds of Tea: the giddineſs was ſometimes ſo conſiderable, as to render it neceſſary for a perſon to attend him, in order to prevent any injury he might ſuffer from falling or other accident. He was bled in the arm freely, ly, but without permanent relief; his complaint returned as ſoon as he was expoſed to his uſual employment. At length he was adviſed to be electrified, and the ſhocks were directed to his head. The next day his pain was diminiſhed, but the day after cloſed the tragical ſcene. I ſaw him a few hours before he died; he was inſenſible; the uſe of his limbs almoſt loſt, and he ſunk very ſuddenly into a fatal apoplexy. Whether the effluvia of the Tea, or electricity was the cauſe of this event is doubtful. In either view the caſe is worthy of attention (*e*).

A young man of a delicate conſtitution, had tried many powerful remedies in vain, for a depreſſion of ſpirits, which he labored under to a degree of melancholy, which rendered his ſituation dangerous to himſelf and thoſe about him. I found he drank Tea very plentifully, and therefore requeſted him to ſubſtitute another kind of diet, which he complied with, and afterwards gradually recovered his uſual health. Some weeks after this, having a large preſent of fine green Tea ſent him, he drank a conſiderable quantity of the infuſion for that and the following day. This was ſucceeded by his former dejection and melancholy, with loſs of memory, tremblings, a proneneſs to great agitation from the moſt trifling circumſtances, and a numerous train of nervous ailments. I ſaw him again, and he immediately attributed his complaints to the Tea he had drank; ſince which he has carefully de-

(*e*) From theſe inſtances of the deleterious effects of Tea, one might be led to ſuppoſe that the ſame unhappy conſequences would frequently attend thoſe who are employed in examining and mixing different kinds of Tea in China; but there the Teas are mixed under an open ſhed, through which the air has a free current, and thereby the odor and the duſt are diſſipated: but in London this buſineſs is uſually done in a back room, confined on every ſide.

nied

nied himſelf the ſame indulgence, and now enjoys his former health.

I have known many other inſtances, where leſs degrees of depreſſion, and other complaints depending upon a relaxed irritable habit, have attended delicate people for many years, and though they have had the advice of ſkilful phyſicians, yet in vain have remedies been adminiſtered, till the patient has refrained from the infuſion of this fragrant exotic.

SECT. IX.

In treating of this ſubſtance, I would not be underſtood to be either a partial advocate, or a paſſionate accuſer. I have often regretted that Tea ſhould be found to poſſeſs any pernicious qualities, as the pleaſure which ariſes from reflecting how many millions of our fellow creatures are enjoying at one hour the ſame amuſing repaſt; the occaſions it furniſhes for agreeable converſation; the innocent parties of both ſexes it daily draws together, and entertains without the aid of ſpirituous liquors; would afford the moſt grateful ſenſations to a ſocial breaſt. But juſtice demands ſomething more. It ſtands charged by many able writers, by public opinion, partly derived from experience, with being the cauſe of many grievous diſorders; all that train of diſtempers included under the name of nervous, are ſaid to be, if not the offspring, at leaſt highly aggravated by the uſe of Tea. To enumerate all theſe, would be to tranſcribe volumes. It is not impoſſible but the charges may be partly true. Let us examine the caſe with all poſſible candor.

The effect of drinking large quantities of any warm aqueous liquor, according to all the experiments we are acquainted with, would be, to enter ſpeedily into the courſe of circulation, and paſs off as ſpeedily by urine or perſpiration, or the encreaſe of ſome of the ſecretions. Its effects on the ſolid parts of the conſtitution would be relaxing, and thereby enfeebling. If this warm aqueous fluid were taken in conſiderable quantities, its effects would be proportionable, and ſtill greater, if it were ſubſtituted inſtead of nutriment.

That all infuſions of herbs, may be conſidered in this light, ſeems not unreaſonable. The infuſion of Tea, nevertheleſs, has theſe two particularities. It is not only poſſeſſed of a ſedative quality (SECT. III. EXP. III. IV.), but alſo of a conſiderable aſtringency (SECT. II. EXP. II.); by which the relaxing power aſcribed to a mere aqueous fluid, is in ſome meaſure corrected. It is on account of the latter, perhaps leſs injurious than many other infuſions of herbs, which, beſides a very ſlight aromatic flavor, have very little if any ſtypticity, to prevent their relaxing debilitating effects.

So far therefore Tea, if not too fine, if not drank too hot, nor in too great quantities, is perhaps preferable to any other vegetable infuſion we know. And if we take into conſideration likewiſe, its known enlivening energy, it will appear that our attachment to Tea, is not merely from its being coſtly or faſhionable, but from its ſuperiority in taſte and effects to moſt other vegetables.

SECT.

SECT. X.

It may be of ſome uſe in our enquiries to conſider its effects where it has been long uſed, and univerſally. Of Japan we know little at preſent: of China we have more recent accounts; from theſe it appears, that Tea of ſome kind, coarſer or finer, is drank by all degrees of people, and copiouſly; that the general proviſion of the lower ranks eſpecially is rice, their beverage Tea. The better kind of people drink Tea, but they live likewiſe on animal food, and live freely.

Of their diſeaſes we know but little, nor what effects Tea may have in this reſpect. They never bleed on any account. The late Dr. Arnot, of Canton, a gentleman who did his profeſſion and his country honor, and was in the higheſt eſtimation with the Chineſe, I am informed was the firſt perſon, who could ever prevail upon any of the Chineſe to be blooded *(f)*, be their maladies what they might. It would appear from hence, that inflammatory diſeaſes were not extremely common; otherwiſe a nation who ſeem ſo fond of life as the Chineſe are reputed to be, would by ſome means or other have admitted of this almoſt only remedy in ſuch caſes. May we infer from hence, that inflammatory diſeaſes are leſs frequent in China, than in ſome other countries, and that probably one cauſe of this may be the conſtant and liberal uſe of this infuſion? perhaps if we take a view of the ſtate of

(f) See Du Halde's hiſtory of China, V. III. p. 362. He obſerves here, that bleeding is not entirely unknown amongſt the Chineſe.

diſeaſes, as exactly deſcribed a century ago, and compare it with what we may obſerve at preſent, we may have a collateral ſupport for this ſuggeſtion. If we conſider the frequency of inflammatory diſeaſes in Sydenham's time, who was both a conſummate judge of theſe diſeaſes, and deſcribed them faithfully, I believe we ſhall find they were then much more frequent than they are at preſent; at leaſt I have been informed ſo by ſome able and obſerving people of the faculty, who moſtly agree, that genuine inflammatory diſeaſes are much more rare at preſent, than they were at the time when Sydenham wrote. It is true, this diſpoſition, admitting it be fact, may ariſe from various cauſes; amongſt the reſt, it is not improbable but Tea may have it's ſhare.

SECT. XI.

Before the uſe of Tea, the general breakfaſt in this country conſiſted of ſomething more ſubſtantial; milk in various ſhapes, ale and beer, with toaſt, cold meat, and other additions. The like additions with ſack, and the moſt generous wines, found their way amongſt the higher orders of mankind. And one cannot ſuppoſe but that ſuch a diet, and the uſual exerciſe they took, would produce a very different ſtate of blood and other animal juices, from that which Tea, a little milk or cream, and bread and butter affords.

It was not the breakfaſt only that ſeems to have contributed its ſhare towards introducing a material alteration in the animal ſyſtem, but the ſubſequent regale likewiſe in the afternoon. Tea is a ſecond time brought before company; it is drank by moſt people, and often in no very ſmall quantities.

tities. Before the introduction of this exotic, it was not unusual to entertain afternoon guests in a very different manner; jellies, tarts, sweetmeats; nay, cold meat, wine, cyder, strong ale, and even spirituous liquors under the title of cordials, were often brought out on these occasions, and perhaps carried to a culpable excess, and much to the injury of individuals.

This kind of repast would tend to keep up the natural inflammatory diathesis, which was the result of vigor, and a plenitude of rich blood; as well as favor diseases originating from such causes. It seems not unreasonable therefore to suppose, that as the diet of our ancestors was more generous, their exercises more athletic, and their diseases more generally, the produce of a rich blood, than are observable in the present times; that these debilitating effects before mentioned may in part be attributed to the use of Tea, as no cause appears to be so universal and so probable.

SECT. XII.

If these suggestions are admitted, they will assist us in determining when and to whom the use of Tea is salutary, and to whom it may be deemed injurious. Those for instance, who either from a natural propensity to generate a rich inflammatory blood, or from exercise or diet, or climate, or all together, are disposed to be in this situation: to these the use of Tea would seem rather beneficial, by relaxing the too rigid solids, and diluting the coagulable lymph of the blood, as a very sensible and ingenious author very justly stiles it *(g)*.

(g) Philosophical Transactions, Vol. LX. 1770. p. 368, & seq.

There

There are idiosyncrases, certain particularities, which are objections to general rules. There are for instance men of this temperament, strong, healthy, vigorous, and with not only the appearance, but the requisites of firm health, to whom a few dishes of Tea would produce the agitations familiar to an hysteric woman: but this is by no means general: in common they bear it well, it refreshes them, they endure fatigue after it, as well as after the most substantial viands. Nothing refreshes them more than Tea, after lasting and vehement exercise. To such it is undoubtedly wholesome, and equal at least, if not preferable, to any other kind of regale now in use.

But if we consider what may reasonably be supposed to happen to those, who are in the opposite extreme of health and vigor, that is, the tender, delicate, enfeebled, whose solids are debilitated, their blood thin and aqueous, the appetite lost or depraved, without exercise, or exercising improperly; in short, where the disposition of the whole frame is altogether opposite to the inflammatory; the free and unrestrained use of this infusion, and such accompaniments, must unavoidably contribute to sink the remains of vital strength still lower.

Between these two extremes there are many gradations; and every thing else being alike, Tea will in general be found more or less beneficial or injurious to individuals, in proportion as their constitutions approach nearer to these opposite extremes. To descend into all the particulars would require experience and abilities, more than I can boast. Suffice it to say, that except as a medicine, or after great fatigue, large quantities are seldom beneficial, nor should it ever be drank very hot; and, as hath been already mentioned, the finer Tea,

Tea, the green eſpecially, is more to be ſuſpected than the common or middling kinds.

S E C T. XIII.

The experiments and obſervations hitherto related, render it evident, that Tea poſſeſſes a fragrant volatile principle, which in general tends to relax and enfeeble the ſyſtem of delicate perſons, particularly when it is drank hot, and in large quantities. I have known many of this frame of conſtitution, who have been perſuaded on account of their health, to deny themſelves this faſhionable infuſion, with great benefit (SECT. VIII.). Others who have found their health impaired by this indulgence, are induced to continue it for want of a proper ſubſtitute, eſpecially for breakfaſt.

But if ſuch cannot wholly omit this favorite regale, they may certainly take it with more ſafety, by boiling the Tea a few minutes, in order to diſſipate this fragrant principle (SECT. III, IV.) which is the moſt noxious; and extract the bitter, aſtringent and moſt ſtomachic part (SECT. II. III.), inſtead of preparing it in the uſual manner by infuſion.

An eminent phyſician in the city, frequently experiencing the prejudicial effects of Tea, by drinking it in the uſual form, was induced from reading a diſſertation upon this ſubject, publiſhed ſome time ſince at Leyden *(b)*, to try the infuſion prepared after another manner. He ordered the Tea to be infuſed in hot water, which after a few hours he cauſed to be poured off, ſtand over night, and to be made warm again in

(b) Siſtens obſervationes ad vires Theæ pertinentes. Lugd. Batav. 1769.

the

the morning for breakfaſt. By this means he aſſures me he can take without inconvenience near double the quantity of Tea, which, when prepared in the uſual method, would formerly have produced many diſagreeable nervous complaints.

The ſame end is obtained by ſubſtituting the extract of Tea (SECT. III. 2.) inſtead of the leaves. I have frequently tried it in the form of Tea, by diſſolving it in warm water, and to me it is a pleaſant ſtomachic bitter; as the fragrancy of the Tea is in this caſe diſſipated, the nervous relaxing effects, which follow the drinking it in the uſual manner, would be in great meaſure avoided. This extract has been imported into Europe from China, in flat round dark colored cakes, not exceeding a quarter of an ounce each in weight, ten grains of which, diſſolved in a ſufficient quantity of water, might ſuffice one perſon for breakfaſt. It might alſo be made here without much expence or trouble (See SECT. III. 2.).

An infuſion of chamomile flowers, or any bitter ſtomachic, taken after drinking Tea, ſometimes prevents the relaxing effects of the foreign herb. The bitter infuſions are alſo more beneficial when drank cool.

It is remarkable, that in all the forms which Du Halde relates, for adminiſtering Tea as a ſtomachic medicine among the Chineſe, it is ordered to be boiled for ſome time, or prepared in ſuch a manner, as to cauſe a diſſipation of its fragrant periſhable flavor, which practice, as it ſeems conſonant to experiments here (SECT. II. III.), may probably have taken its riſe in China, from long experience and repeated facts.

SECT.

SECT. XIV.

Perhaps it will be deemed rather foreign to an eſſay upon this ſubject, to take a conciſe view of the manners, or morals if the reader pleaſes, of the Chineſe, as we have done of their diſeaſes; but as thoſe who are beſt acquainted with human nature, ſeem to aſcribe even to their food, and way of life, as well as to their climate and education, certain propenſities at leaſt to vice and virtue, it may be of uſe to draw what light we can in theſe reſpects, from the character of a people, who have uſed the infuſion of Tea for ſucceſſive generations.

They are in general deſcribed to be a people of moderate ſtrength of body, not capable of much hard labor, rather feeble when compared with the inhabitants of ſome nations, excelling in ſome minute fabricks and manufactures, but exhibiting no proofs of elevated genius in architecture, either civil or military. They are deſcribed to be puſillanimous, cunning, extremely libidinous, and remarkable for diſſimulation and ſelfiſhneſs *(i)*, effeminate, revengeful and diſhoneſt *(k)*.

It would be unjuſt to aſcribe all theſe qualities to their manner of living: other cauſes have undoubtedly their ſhare: but it may be ſuſpected, that the manner of life, or kind of diet, that tends to debilitate, virtually contributes to the encreaſe of the meaner qualities. Where force of body is wanting, cunning often ſupplies its place; and if not regu-

(i) See Anſon's voyage round the world, 8vo. p. 366, and many later authorities.

(k) See likewiſe Du Halde's hiſtory of China, Vol. II. p. 75, 130 et ſeq.

lated by other principles, it would diſcover its effects more univerſally; and thus will take place whether the debility is natural, or acquired by a diet that enfeebles the body. That there is a probity, fortitude, generoſity in female minds, not inferior to the like qualities poſſeſſed by the other ſex, is moſt certain, but that it is generally ſo, may perhaps be doubted.

Whether the preſent age exhibits as many inſtances of ſuperior excellence as the preceding, is beyond my abilities to determine: that it is tarniſhed more than ſome others with one vice at leaſt, ſeems generally confeſſed: and it may perhaps be a problem not unworthy conſideration, whether the general uſe of Tea, may not gradually encreaſe the diſpoſition. For whatever tends to debilitate, ſeems for the moſt part to encreaſe corporeal ſenſibility. The ſame perſon who in health, does not ſtart at the firing of a cannon, ſhall be extremely diſconcerted, when ſunk by diſeaſe to the border of effeminacy, at the ſudden opening of a door. Deſire is not always proportioned to bodily ſtrength: it may be ſtrongeſt when the corporeal ſtrength is at the loweſt ebb; it is often found ſo; and therefore another reaſon occurs, why the general uſe of Tea ought not to be conſidered, as amongſt the moſt indifferent of all ſubjects.

S E C T. XV.

From what has been ſaid upon this ſubject, it will probably appear, that children and very young perſons in general, ſhould as much as poſſible be deterred from the uſe of this infuſion. It weakens their ſtomachs, impairs the digeſtive powers, and favors the generation of many diſeaſes. We ſeldom perceive the rudiments of ſcrophulous diſeaſes, ſo often

as

as in the weak feeble offspring of the inhabitants of towns; and whoſe breakfaſt and ſupper often conſiſts of the weak runnings of ordinary Tea, with its uſual appurtenances. In better families experience has directed to a better choice; amongſt many it loſes ground, from a knowledge of its injurious effects. It ought by no means to be the common diet of boarding ſchools; if it be allowed ſometimes as a treat, they ſhould at the ſame time be informed, that the conſtant uſe of it would be injurious to their health, ſtrength, and conſtitution in general.

S E C T. XVI.

Thus far I have chiefly endeavoured to trace the effects of Tea as a part of our diet. In medicine it has at preſent but very little reputation amongſt us. It is even ſcarcely ever recommended as a part of the furniture of a ſick chamber; it is ſeldom mentioned even as a gentle diaphoretic: in caſes, however, where it is neceſſary to dilute and relax, to promote the thinner ſecretions, it at leaſt promiſes as much advantage as moſt other infuſions. For beſides its other effects, it ſeems to contain ſomething ſedative in its compoſition (Sect. III. Exp. III. IV.), not altogether unlike an opiate; like this claſs of medicines, it mitigates uneaſineſs, perhaps more than any other merely aqueous infuſion: and like very ſmall doſes of opium, it ſometimes prevents reſt, and gives a temporary flutter to the ſpirits.

Where therefore large quantities of the infuſion muſt be taken, to produce or ſupport a conſiderable diaphoreſis, a decoction of Tea, or a ſtrong infuſion, may be adminiſtered with great propriety, particularly in inflammatory complaints;

the ſedative power of Tea, aſſiſted by the diluting effects of warm water, generally producing a diaphoreſis, without ſtimulating the ſyſtem. The Chineſe moſt commonly give it as a medicine in decoction, in a variety of diſeaſes: but if the infuſion were drawn from a large proportion of fine Tea, and ſoon poured off, that the fineſt part may be procured, and drank warm, it would ſeem preferable as an attenuant and relaxant.

I have more than once given fine green Tea in ſubſtance with ſome diluting vehicle, and obſerved the ſame effects nearly, as are produced from taking the infuſion. Thirty grains of this kind of Tea powdered, taken three or four times, at as many hours interval, generally relaxes the ſolids, diminiſhes heat and reſtleſſneſs, and induces perſpiration. Such a doſe as produces a ſlight nauſea, which this quantity uſually does, more certainly induces a perſpiration, and a mitigation of the ſymptoms accompanying inflammatory complaints. If this doſe be doubled, the nauſea and ſickneſs are encreaſed, and a diſagreeable pain or load is felt for ſome time about the region of the ſtomach, which uſually goes off with a laxative ſtool.

S E C T. XVII.

It is ſaid that in Japan and China, the ſtone is a very unuſual diſtemper, and the natives ſuppoſe that Tea has the quality to prevent it. So far as it ſoftens and meliorates the water, it may certainly be of uſe *(l)*. We may alſo obſerve

(l) By long boiling, water is certainly freed from ſome of the earthy, and ſaline ubſtances it may contain, and thereby rendered conſiderably ſofter; but it is by no means altered in theſe reſpects by infuſing with Tea. See Percival's experiments and obſervations on water, p. 27 et 33.

here,

here, that every ſolvent is capable of taking up a limited quantity only of the ſolvend, and when fully ſaturated with it, is incapable of ſuſpending it long; hence it is plain, that the quantity of the ſtony matter carried off, muſt be greater when the urine is encreaſed in quantity, and has not been too long retained in the bladder: and therefore as Tea is diuretic, it may in this view prove lithonthriptic.

Tea, we have already obſerved, contains an aſtringent antiſeptic quality (SECT. II. EXP. I. II.). It likewiſe poſſeſſes no inconſiderable degree of bitterneſs; and as the uvæ urſi, and other bitters have mitigated ſevere paroxyſms of the ſtone, may not Tea prove ſerviceable alſo by its antacid quality?

It is an obſervation I have often had occaſion to make, that people after violent exerciſe, or coming off a journey much fatigued, and affected with a ſenſe of general uneaſineſs, attended with thirſt and great heat; by drinking a few cups of warm Tea, have generally experienced immediate refreſhment. It alſo proves a grateful diluent, and agreeable ſedative, after a full meal, when the ſtomach is oppreſſed, the head pained, and the pulſe beats high *(m)*.

SECT. XVIII.

I ſhall finiſh theſe remarks with ſome reflections on this herb, conſidered in another light.

As luxury of every kind has augmented in proportion to the encreaſe of foreign ſuperfluities, it has contributed more or leſs its ſhare towards the production of thoſe low nervous diſeaſes,

(m) Le Compte's memoirs and obſervations, p. 227. Home's Principia Medicinæ, p. 5. Percival's experimental eſſays, p. 130. See alſo Tiſſot on the diſeaſes of literary and ſedentary perſons, p. 145.

which

which are now ſo frequent. Amongſt theſe cauſes, exceſs in ſpirituous liquors is one of the moſt conſiderable; but the firſt riſe of this pernicious cuſtom is often owing to the weakneſs and debility of the ſyſtem, brought on by the daily habit of drinking Tea *(n)*; the trembling hand ſeeks a temporary relief in ſome cordial, in order to refreſh and excite again the enfeebled ſyſtem; whereby ſuch almoſt by neceſſity fall into a habit of intemperance, and frequently intail upon their offspring, a variety of diſtempers, which otherwiſe would not probably have occurred.

S E C T. XIX.

Another bad conſequence reſulting from the univerſal cuſtom of Tea drinking, particularly affects the poor laboring people, whoſe daily earnings are ſcanty enough to procure them the neceſſary conveniences of life, and wholeſome diet. Many of theſe, too deſirous of vieing with their ſuperiors, and imitating their luxuries, throw away their little earnings upon this faſhionable herb, and are thereby inconſiderately deprived of the means to purchaſe proper wholeſome food for themſelves and their families.

I have known ſeveral miſerable families thus infatuated, their emaciated children laboring under various ailments depending upon indigeſtion, debility, and relaxation. Some at length have been ſo enfeebled, that their limbs have become diſtorted, their countenance pale, and a maraſmus has cloſed the tragedy.

(n) See Percival's experimental eſſays, p. 126.

Theſe

Theſe effects are not to be attributed ſo much to the peculiar properties of this coſtly vegetable, as to the want of proper food, which the expence of the former deprived theſe poor people from procuring. I knew a family of this ſtamp, conſiſting of a mother and ſeveral children, whoſe fondneſs for Tea was ſo great, and their earnings ſo ſmall, that three times a day, as often as their meals, which generally conſiſted of the ſame articles, they regularly ſent for Tea and ſugar, with a morſel of bread to ſupport nature; by which practice they daily grew more enfeebled; thin emaciated habits and weak conſtitutions characteriſed this diſtreſſed family, till ſome of the children were removed from this baneful nurſery, who afterwards acquired tolerable health.

An ingenious author obſerves that as much ſuperfluous money is expended on Tea and ſugar in this kingdom, as would maintain four millions more of ſubjects in bread *(o)*. And the author of the farmers letters calculates that the entertainment of ſipping Tea coſts the poor each time as follows:

	d.
The tea - - - - - - -	¾
The ſugar - - - - - - -	½
The butter - - - - - -	1
The fuel and wear of the Tea equipage - -	¼
	2½

When Tea is drank twice a day, the annual expence amounts to 7l. 12s. a head; and the ſame judicious writer eſtimates the bread neceſſary for a laborer's family of five perſons, at 14l. 15s. 9d. per annum *(p)*, by which it appears, that the yearly expence of Tea, ſugar, &c. for two perſons, exceeds

(o) Eſſays on huſbandry, p. 166
(p) Vol. I. p. 202 and 299.

that

that of the neceſſary article of bread, ſufficient for a family of five perſons.

S E C T. XX.

It appears alſo from a moderate calculation, that three million pounds of Tea are annually conſumed in England ; and domeſtic experience teaches us, that with each pound of Tea, ten pounds of butter at leaſt are conſumed. Hence the conſumption of butter with this injurious aliment, if aliment it may be called, amounts annually to the amazing quantity of thirty millions of pounds. It is likewiſe to be premiſed, that at leaſt five gallons of milk, are neceſſary to procure one pound of butter *(q)*. This being granted, we may conclude farther:

Suppoſe one gallon of milk with bread, would ſuffice three laboring people for breakfaſt and ſupper, and that theſe meals conſtitute half of their food, it follows, that from this faſhionable cuſtom of Tea drinking, this kingdom cannot ſupply food for ſo many people as it otherwiſe could, were the inhabitants to live in a more ſimple manner, by at leaſt one million. But ſuppoſing we allow half a million for the bread eaten with the milk, and for the uſes of the milk after the butter has been taken from it, the deficiency ſtill amounts to the amazing number of half a million of people !

(q) Compare Halleri Elem. Phyſ. T. 7. P. 11. p. 33, 34.

F I N I S.

Printed in the United States
By Bookmasters